Intermediate 2
PHYSICS

Andrew McCormick
Arthur Baillie

Hodder & Stoughton

A MEMBER OF THE HODDER HEADLINE GROUP

Acknowledgements

The publishers would like to thank the following individuals, institutions and companies for permission to reproduce photographs in this book. Every effort has been made to trace ownership of copyright. The publishers are happy to make arrangements with any copyright holder whom it has not been possible to contact.

Alex Bartel/ Science Photo Library (10.9); Debby Besford/ Science Photo Library (6.5b); Martin Bond/ Science Photo Library (15.1); Martin Dohrn/ Science Photo Library (12.4); Adam Hart-Davis/ Science Photo Library (5.5); Sergio Dorantes/CORBIS (6.2); James King Holmes/ Science Photo Library (12.6); Larry Lee Photography/CORBIS (7.9); LifeFile (6.11); Vaughan Melzer/JVZ/ Science Photo Library (15.2); Will & Deni McIntyre/ Science Photo Library (10.3); David Parker/Science Photo Library (5.4, 10.6, 15.6); NPG, London (6.5a); PA Photos (9.1, 11.12); PhilipHarris Education (6.1); Rosenfeld Images Ltd/ Science Photo Library (6.3); Josh Sher/Science Photo Library (14.3); Takeshi Takahara/ Science Photo Library (6.4)

The renogram images (Figure 12.7) were kindly provided by P.S.Cosgriff, Consultant Medical Physicist, Pilgrim Hospital, Boston, Lincolnshire.

The illustrations were drawn by Jeff Edwards and Chartwell Illustrators

Sample examination questions reproduced with kind permission from SQA

Orders: please contact Bookpoint Ltd, 130 Milton Park, Abingdon, Oxon OX14 48B. Telephone: (44) 01235 827720. Fax: (44) 01235 400454. Lines are open from 9.00 - 6.00, Monday to Saturday, with a 24 hour message answering service. Email address: orders@bookpoint.co.uk

British Library Cataloguing in Publication Data
A catalogue record for this title is available from the British Library

Intermediate 2 Physics: Without Answers
ISBN 0 340 78269 2

Published by Hodder & Stoughton Educational
First Published 2001
Impression number   10 9 8 7 6 5 4 3 2 1
Year   2007 2006 2005 2004 2003 2002 2001

Intermediate 2 Physics: With Answers
ISBN 0 340 78268 4

Published by Hodder & Stoughton Educational
First Published 2001
Impression number   10 9 8 7 6 5 4 3 2 1
Year   2007 2006 2005 2004 2003 2002 2001

Cover photo from David Ducross/Science Photo Library
Typeset by Wyvern 21 Limited
Printed in Great Britain for Hodder & Stoughton Educational, a division of Hodder Headline Plc, 338 Euston Road, London NW1 3BH by The Bath Press Limited

Section 1
Electricity and Electronics

Section 2
Mechanics and Heat

Section 3
Waves and Optics

Section 4
Radioactivity

There is no prescribed route to follow to arrive at a new idea. You have to make the intuitive leap. But the difference is that once you have made that intuitive leap you have to justify it by filling in the intermediate steps

Stephen Hawking Lucasian Professor of Mathematics Cambridge University and author of 'A Brief History of Time'

Ye cannae change the laws of physics Captain

Scotty in Star Trek

This book was written to fulfil the main requirements of the Intermediate 2 Physics course. This has taken us a great deal of time and effort and we are thankful to friends, colleagues and particularly our families for their patience in this task. We are especially grateful to Roddy Glen for yet again providing editorial advice. His wise words, suggestions, corrections and help have helped us to fulfil a very difficult task. Any remaining omissions and errors are due to us. We would also like to thank Stephen Halder at Hodder and Stoughton for his help, advice and pressure to complete the task. Finally it is our hope that not only will this book help students prepare for and pass the examination but will also provide some sense of enjoyment and stimulate students' interest in physics. They may even think physics is fun.

Andrew McCormick
Arthur Baillie
March 2001

For each chapter you should be able to:

1 Use SI units of all quantities in each section.

2 Give answers to calculations, to an appropriate number of significant figures.

3 Check answers to calculations.

4 Use prefixes (μ, m, k, M).

5 Use scientific notation.

Electricity and Electronics

1 Circuits

At the end of this chapter you should be able to:

1 State that electrons are free to move in a conductor.

2 Describe the electrical current in terms of the movement of charges around a circuit.

3 Carry out calculations involving $Q = It$.

4 Distinguish between conductors and insulators and give examples of each.

5 Draw and identify circuit symbols for an ammeter, voltmeter, battery, resistor, variable resistor, fuse, switch and lamp.

6 State that the voltage of a supply is a measure of the energy given to the charges in a circuit.

7 State that an increase in the resistance of a circuit leads to a decrease in the current in that circuit.

8 Draw circuit diagrams to show the correct positions of an ammeter and voltmeter in a circuit.

9 State that in a series circuit the current is the same at all positions.

10 State that the sum of the potential differences across the components in series is equal to the voltage of the supply.

11 State that the sum of the currents in parallel branches is equal to the current drawn from the supply.

12 State that the potential difference across components in parallel is the same for each component.

13 State that V/I for a resistor remains approximately constant for different currents.

14 Carry out calculations involving the relationship $V = IR$.

15 Carry out calculations involving the relationships $R_T = R_1 + R_2 + R_3$ and $1/R_T = 1/R_1 + 1/R_2 + 1/R_3$.

16 State that a potential divider consists of a number of resistors, or a variable resistor, connected across a supply.

17 Carry out calculations involving potential differences and resistances in a potential divider.

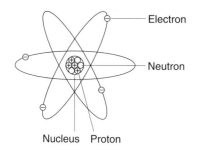

Figure 1.1 ▲
Model of an atom

What is electricity?

All solids, liquids and gases are made up of atoms. An atom consists of a positively charged centre or nucleus surrounded by a 'cloud' of rapidly revolving negative charges called electrons. The nucleus is made of particles called protons (positively charged) and neutrons (uncharged).

Charge is measured in coulombs (C). The charge on a proton is 1.6×10^{-19} C. The charge on an electron is the same size as the charge on a proton.

Charge, current and time

Consider a simple electrical circuit – a lamp connected to a battery. The lamp lights up. This is due to electrons from the negative terminal of the battery moving through the wires and lamp to the positive terminal of the battery. This movement of negative charges is called an electric current (or current for short). A current is a movement of electrons. Therefore, when there is a current, (negative) charge is transferred.

The amount of charge transferred is given by:

$$\text{charge transferred} = \text{current} \times \text{time}$$

$$Q = It$$

where Q = charge transferred, I = current, t = time.

Current is measured in amperes (A) and time in seconds (s).

1 coulomb = 1 ampere second (1 C = 1 A s)

Example
How much charge is transferred in 5 minutes by a current of 0.2 A?

Solution

$Q = It$

$Q = 0.2 \times (5 \times 60)$

$Q = 60$ C

Conductors and insulators

Electrons can only move from the negative terminal to the positive terminal of a battery if there is an electrical path between them. Materials which allow electrons to move through them easily, to form an electric current, are known as conductors. Conductors are mainly metals, such as copper, gold and silver. However, carbon is also a good conductor.

Materials which do not allow electrons to move through them easily are called insulators. Glass, plastic, wood and air are examples of insulators.

Voltage or potential difference

A battery changes chemical energy into electrical energy. This electrical energy is carried by the electrons that move round the circuit and converted into other forms of energy, e.g. heat and light, by components in the circuit such as a lamp. The amount of electrical energy the electrons have at any point in a circuit is known as their 'potential'. As electrons move between two points in an electric circuit, they transfer electrical energy into other forms of energy. This means that the electrons have a different amount of electrical energy (or potential) at the two points. There is a potential difference (or p.d.) between the points.

The electrical energy given to the electrons by the battery is a measure of the voltage or potential difference between the two terminals of the battery. To be exact, the voltage or p.d. of a battery is the electrical energy given to one coulomb of charge passing through the battery. For example a 6-volt battery gives four times as much energy to each coulomb of charge passing through it compared to a 1.5-volt battery.

Circuit symbols

The circuit symbols for a battery, resistor, variable resistor, fuse, switch, lamp, ammeter and voltmeter are shown in figure 1.2.

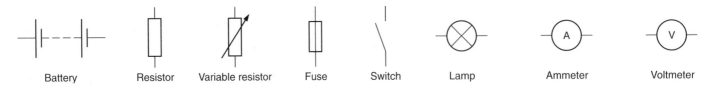

Battery   Resistor   Variable resistor   Fuse   Switch   Lamp   Ammeter   Voltmeter

Figure 1.2 ▲
Common symbols

Measuring current

An ammeter is used to measure electric current. It is measured in amperes (A). Figure 1.3 shows how an ammeter is connected in an electrical circuit.

Figure 1.3 ▶
Connecting an ammeter into an electrical circuit

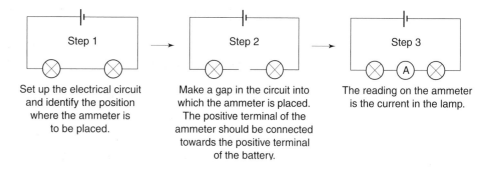

Set up the electrical circuit and identify the position where the ammeter is to be placed.

Make a gap in the circuit into which the ammeter is placed. The positive terminal of the ammeter should be connected towards the positive terminal of the battery.

The reading on the ammeter is the current in the lamp.

An ammeter measures the current (i.e. 'counts' the number of coulombs each second) through a component.

Measuring voltage

A voltmeter is used to measure voltage or potential difference (p.d.). It is measured in volts (V). Figure 1.4 shows how a voltmeter is connected to an electrical circuit.

Figure 1.4 ▶
Connecting a voltmeter into an electrical circuit

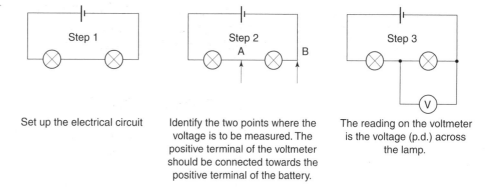

Set up the electrical circuit

Identify the two points where the voltage is to be measured. The positive terminal of the voltmeter should be connected towards the positive terminal of the battery.

The reading on the voltmeter is the voltage (p.d.) across the lamp.

A voltmeter measures the voltage or p.d. across a component (i.e. the number of joules of energy transferred by each coulomb of charge).

Ammeters and voltmeters can be connected to the same circuit using the instructions given above.

Resistance

All materials oppose current passing through them. This opposition to the current is called resistance. Resistance is measured in ohms (Ω).

For most materials resistance depends on the:

- Type of material – the better the conductor the lower the resistance.
- Length of the material – the longer the material the higher the resistance.
- Thickness of the material – the thinner the material the higher the resistance.
- Temperature of the material – the higher the temperature the higher the resistance.

For a resistor, the ratio of $\dfrac{\text{voltage (p.d.) across resistor}}{\text{current through resistor}}$

remains approximately constant.

This constant is the resistance of the resistor. Therefore, the resistance of a resistor can be calculated provided the voltage or potential difference across the resistor and the current through the resistor are known. Then:

$$\text{resistance} = \frac{\text{voltage (p.d.) across resistor}}{\text{current through resistor}} \quad \text{i.e. } R = \frac{V}{I}$$

This is known as Ohm's law.

In units: $\text{ohms } (\Omega) = \dfrac{\text{volts (V)}}{\text{amperes (A)}}$

Ohm's law is normally written as

Voltage (p.d.) across resistor = current through resistor × resistance of resistor

or $V = IR$

Example

There is a current of 1.5 A in a lamp. The lamp has a resistance of 8 Ω. Calculate the potential difference across the ends of the lamp.

Solution

$V = IR$

$V = 1.5 \times 8$

$V = 12 \text{ V}$

A resistor whose resistance can be changed is known as a variable resistor. The resistance is changed by altering the length of the wire in the resistor (the longer the wire, the higher the resistance) – the higher the resistance the smaller the current. Variable resistors are often used as volume or brightness controls on televisions, and dimmers on lights.

Types of circuit

Electrical components, such as lamps and resistors, can be connected in series, in parallel or a mixture of series and parallel. A series circuit has only one electrical path from the negative terminal of the battery to the positive terminal. A parallel circuit has more than one electrical path from the negative terminal of the battery to the positive terminal. Figure 1.5 shows three lamps connected in series, while figure 1.6 shows three lamps connected in parallel. Figure 1.7 shows a mixed series and parallel circuit in which a lamp is connected in series with two resistors which are connected in parallel.

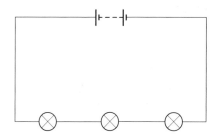

Figure 1.5 ▲
A series circuit

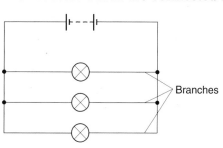

Branches

Figure 1.6 ▲
A parallel circuit

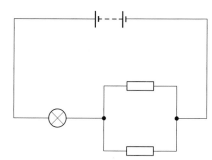

Figure 1.7 ▲
A mixed series and parallel circuit

A series circuit

Three resistors of value 1 Ω, 4 Ω and 3 Ω are connected in series as shown in figure 1.8. Ammeters are connected to measure the current at various positions. Voltmeters have also been connected to measure the voltage (p.d.) across each resistor. The table shows the readings on the meters and the values of the resistances calculated using Ohm's law.

Figure 1.8 ▶
Measuring the current through and the voltage (p.d.) across resistors connected in series

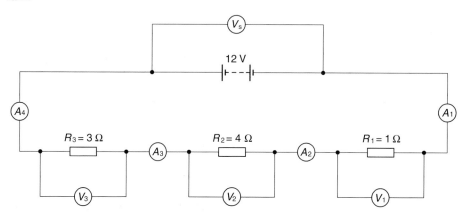

Current	Voltage	Resistance
$A_1 = 1.5$ A	$V_1 = 1.5$ V	$R_1 = 1\ \Omega$
$A_2 = 1.5$ A	$V_2 = 6.0$ V	$R_2 = 4\ \Omega$
$A_3 = 1.5$ A	$V_3 = 4.5$ V	$R_3 = 3\ \Omega$
$A_4 = 1.5$ A		
–	$V_s = 12$ V	–

From the table we can conclude that:

- The current at different positions in a series circuit is the same.
- The sum of the voltages (p.d.'s) across the resistors is equal to the supply voltage.

Figure 1.9 shows a single resistor connected to the same supply voltage. This circuit must have the same combined or total resistance as figure 1.8 since it has the same supply voltage and the same current. Using Ohm's law on this circuit gives a value for the single resistor of 8 Ω (as $R = V_s/I = 12/1.5 = 8\ \Omega$).

Figure 1.9 ▶

Identical circuit to that shown in figure 1.8

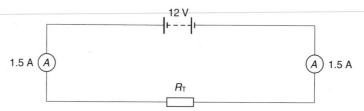

Comparing figures 1.8 and 1.9 we see that the combined or total resistance of a number of resistors connected in series is equal to the sum of the individual resistances.

For the series circuit shown in figure 1.10 we have:

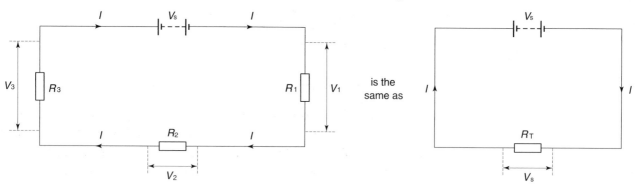

Figure 1.10 ▲

Identical circuits provided $R_T = R_1 + R_2 + R_3$

- The current is the same at all positions – current does not split up.
- The supply voltage is equal to the sum of the voltages (potential differences) across the components, i.e. $V_s = V_1 + V_2 + V_3$
- The total resistance (R_T) of the circuit is equal to the sum of the individual resistances, $R_T = R_1 + R_2 + R_3$.

Example

Three resistors of value 100 Ω, 47 Ω and 33 Ω are connected in series. What is the total resistance of the resistors?

Solution

$$R_T = R_1 + R_2 + R_3$$
$$= 100 + 47 + 33$$
$$= 180 \ \Omega$$

Example

For the circuit shown in figure 1.11, calculate the readings on the ammeters A_1, A_2 and the potential difference across each of the resistors.

Solution

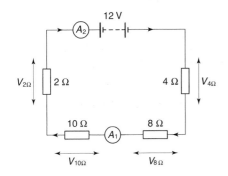

Figure 1.11 ▲
Example of a series circuit

Total resistance of circuit $= R_T = R_1 + R_2 + R_3 + R_4$
$$= 4 + 8 + 10 + 2 = 24 \ \Omega$$

From Ohm's law: Circuit current $= I = \dfrac{V_s}{R_T} = \dfrac{12}{24} = 0.5 \ \text{A}$

$A_1 = A_2 = 0.5$ A (since current in series circuit is the same at all points)

The potential differences across each resistor are calculated using Ohm's law:

Voltage (p.d.) across resistor = current through resistor × resistance of resistor

$$V_{4\Omega} = I \times R_{4\Omega} = 0.5 \times 4 = 2 \ \text{V}$$
$$V_{8\Omega} = I \times R_{8\Omega} = 0.5 \times 8 = 4 \ \text{V}$$
$$V_{10\Omega} = I \times R_{10\Omega} = 0.5 \times 10 = 5 \ \text{V}$$
$$V_{2\Omega} = I \times R_{2\Omega} = 0.5 \times 2 = 1 \ \text{V}$$

A parallel circuit

Two resistors of value 6 Ω and 12 Ω are connected in parallel as shown in figure 1.12. Ammeters are connected to measure the current at various positions. Voltmeters have also been connected to measure the potential difference across each resistor. The table shows the readings on the meters and the values of the resistances calculated using Ohm's law.

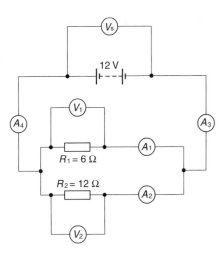

Figure 1.12 ▲
Measuring the current through and the voltage (p.d.) across resistors connected in parallel

	Current	Voltage	Resistance
	$A_1 = 2$ A	$V_1 = 12$ V	6 Ω
	$A_2 = 1$ A	$V_2 = 12$ V	12 Ω
	$A_3 = 3$ A	–	–
	$A_4 = 3$ A		
	–	$V_s = 12$ V	–

From the table we can conclude that:

● The circuit current for a parallel circuit is equal to the sum of the currents in the branches.
● The voltages (p.d.'s) across the resistors connected in parallel are the same.

Figure 1.13 shows a single resistor connected to the same supply voltage. This circuit must have the same combined or total resistance as figure 1.12 since it has the same supply voltage and the same current from the supply. Using Ohm's law on this circuit gives a value for the single resistor of 4 Ω (as $R = V_s/I = 12/3 = 4$ Ω).

Comparing figures 1.12 and 1.13 we see that the combined resistance of a number of resistors connected in parallel is smaller than any of the individual resistances.

The combined resistance of 4 Ω is obtained for these resistors as follows:

$$\frac{1}{R_T} = \frac{1}{R_1} + \frac{1}{R_2} = \frac{1}{6} + \frac{1}{12} = 0.167 + 0.083 = 0.25$$

$$\frac{1}{R_T} = 0.25$$

$$R_T = \frac{1}{0.25} = 4 \ \Omega$$

For the parallel circuit shown in figure 1.14 we have:

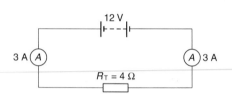

Figure 1.13 ▲
Identical circuit to that shown in figure 1.12

Figure 1.14 ▶
Identical circuits provided $\frac{1}{R_T} = \frac{1}{R_1} + \frac{1}{R_2}$

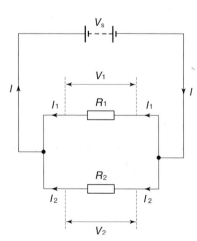

is the same as

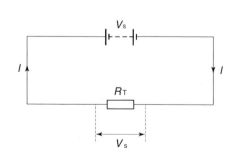

- Current splits up, circuit current = sum of currents in the branches i.e. $I = I_1 + I_2$.
- The voltage (p.d.) across resistors connected in parallel is the same, i.e. $V_1 = V_2$ and in this case $V_s = V_1 = V_2$.
- The total resistance of the circuit is found using $\dfrac{1}{R_T} = \dfrac{1}{R_1} + \dfrac{1}{R_2}$.

Note. For a parallel circuit, the total resistance is less than the value of the smallest resistance connected in parallel.

Example
Three resistors of resistance 10 Ω, 20 Ω and 10 Ω are connected in parallel. Calculate the total resistance of the resistors.

Solution

Figure 1.15 ▶
Example of a parallel circuit

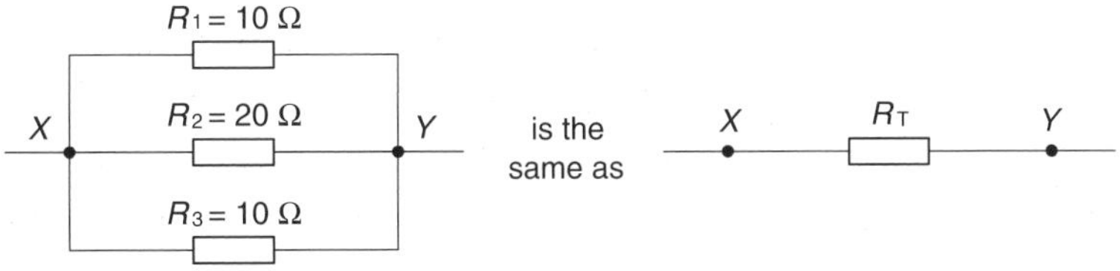

$$\frac{1}{R_T} = \frac{1}{R_1} + \frac{1}{R_2} + \frac{1}{R_3} = \frac{1}{10} + \frac{1}{20} + \frac{1}{10} = 0.1 + 0.05 + 0.1 = 0.25$$

$$\frac{1}{R_T} = 0.25$$

$$R_T = \frac{1}{0.25} = 4 \ \Omega$$

As a check on our answer we would expect it to be smaller than 10 Ω.

Example
Two resistors each of resistance 1 kΩ are connected in parallel. Calculate their total resistance.

Solution

$$\frac{1}{R_T} = \frac{1}{R_1} + \frac{1}{R_2} = \frac{1}{1000} + \frac{1}{1000} = 0.001 + 0.001$$

$$\frac{1}{R_T} = 0.002$$

$$R_T = \frac{1}{0.002} = 500 \ \Omega$$

Note that the total resistance of two identical resistors connected in parallel is half that of one of the resistors.

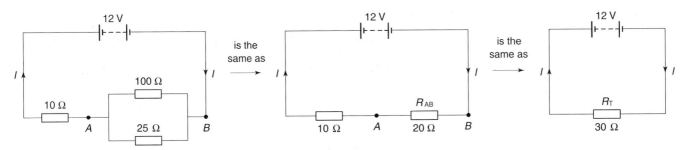

Figure 1.16 ▲
Example of a mixed series and parallel circuit

Example

In the circuit shown in figure 1.16, calculate (a) the total resistance, (b) the current drawn from the battery, (c) the p.d. across the 10 Ω resistor.

Solution

(a) $\dfrac{1}{R_{AB}} = \dfrac{1}{R_1} + \dfrac{1}{R_2} = \dfrac{1}{100} + \dfrac{1}{25} = 0.01 + 0.04 = 0.05$

$R_{AB} = \dfrac{1}{0.05} = 20\ \Omega$

Total resistance $R_T = 10 + R_{AB} = 10 + 20 = 30\ \Omega$

(b) $V_S = I\,R_T$ gives $12 = I \times 30$ Hence $I = \dfrac{12}{30} = 0.4$ A

(c) Voltage (p.d.) across 10 Ω resistor = current through resistor × resistance of resistor

$V = I\,R = 0.4 \times 10 = 4$ V

Potential dividers

For a series circuit, the supply voltage is equal to the sum of the potential differences (voltages) across the individual resistors, i.e. $V_s = V_1 + V_2$. This means that the supply voltage is split up into, in this case, two smaller bits, and so the supply voltage is divided between the components.

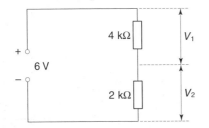

Figure 1.17 ▲
A potential divider circuit

A potential divider consists of two devices, usually resistors, connected in series as shown in figure 1.17. The p.d. of the supply is divided up into two smaller p.d.'s.

$R_T = R_1 + R_2$

$R_T = 2000 + 4000 = 6000\ \Omega$

$I_{circuit} = \dfrac{V_S}{R_T} = \dfrac{6}{6000} = 0.001$ A

$V_1 = I_{circuit} \times R_1 = 0.001 \times 2000 = 2$ V

$V_2 = I_{circuit} \times R_2 = 0.001 \times 4000 = 4$ V

Note. The values of V_1 and V_2 depend on the values of R_1 and R_2.

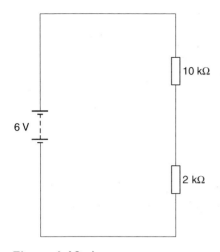

Figure 1.18 ▲
Example of a potential divider

Example

A student sets up an electrical circuit as shown in figure 1.18.

Calculate the potential difference across the 2 kΩ resistor.

Solution

$$R_T = R_1 + R_2$$

$$R_T = 10\ 000 + 2000 = 12\ 000\ \Omega$$

$$I_{circuit} = \frac{V_s}{R_T} = \frac{6}{12\ 000} = 5 \times 10^{-4}\ A$$

$$V_{2\ k\Omega} = I_{circuit} \times R_{2\ k\Omega} = 5 \times 10^{-4} \times 2000 = 1.0\ V$$

Physics facts and key equations for circuits

- Charge is measured in coulombs (C), current in amperes (A) and time in seconds (s).

- Charge = current × time, i.e. $Q = It$.

- The voltage of a supply is a measure of the energy given to one coulomb of charge.

- An ammeter is connected in series.

- A voltmeter is connected in parallel.

- Voltage (or p.d.) is measured in volts (V), current in amperes (A) and resistance in ohms (Ω).

- Voltage (p.d.) across a resistor = current through resistor × resistance of resistor, i.e. $V = IR$ – this is known as Ohm's law.

- The resistance of a resistor remains constant for different currents provided the temperature of the resistor does not change

- In a series circuit:

  – the current is the same at all points, i.e.
  $$I_1 = I_2 = I_3$$

  – the supply voltage is equal to the sum of the voltages (p.d.'s) across components, i.e.
  $$V_s = V_1 + V_2 + V_3$$

  – the total (or combined) resistance is equal to the sum of the individual resistors i.e.,
  $$R_T = R_1 + R_2 + R_3$$

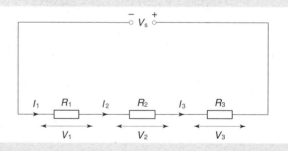

Figure 1.19

- In a parallel circuit:

  – the circuit current is equal to the sum of the currents in the branches, i.e.
  $$I_C = I_1 + I_2 + I_3$$

  – the voltage (p.d.) across components is the same, i.e. $V_S = V_1 = V_2 = V_3$

continued ➤

Physics facts and key equations for circuits *continued*

– the total (or combined) resistance is given

by $\dfrac{1}{R_T} = \dfrac{1}{R_1} + \dfrac{1}{R_2} + \dfrac{1}{R_3}$

– the total resistance is less than the resistance of the smallest resistor.

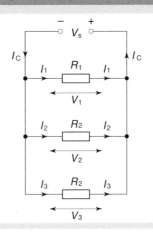

Figure 1.20

End-of-Chapter Questions

1. When a kettle is switched on, there is a current of 10 A in the element. The kettle is switched on for 2 minutes. How much charge flows through the heating element in this time?

2. Calculate the time taken for a current of 2.0 A to transfer 500 C of charge through a lamp.

3. 150 C of charge pass through a resistor in 5 minutes. What is the current in the resistor?

4. An electronic game works from a 9 V supply. What does a supply voltage of 9 V mean?

5. Draw the circuit symbol for **(a)** a battery, **(b)** a lamp, **(c)** a switch, **(d)** a resistor, **(e)** a variable resistor, **(f)** a fuse, **(g)** an ammeter and **(h)** a voltmeter.

6. In the circuits shown in figure 1.21, what are the readings on **(a)** ammeters A_1, A_2 and A_3, **(b)** voltmeters V_1, V_2, and V_3?

7. Three resistors of value 47 Ω, 100 Ω and 150 Ω are connected in series. Find the total resistance of these three resistors.

8. Three resistors of value 20 Ω, 20 Ω and 10 Ω are connected in parallel. Find the total resistance of these three resistors.

9. Four resistors are arranged as shown in figure 1.22. Calculate the resistance between X and Y.

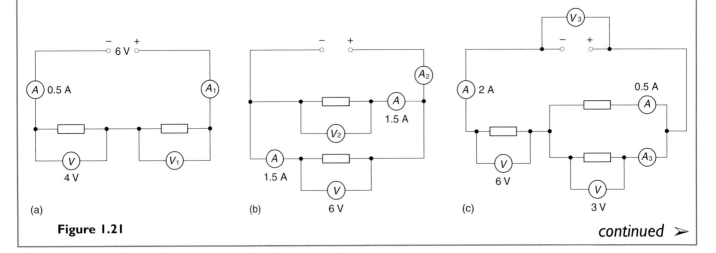

(a) (b) (c)

Figure 1.21

continued ➤

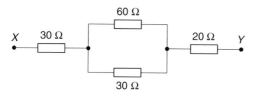

Figure 1.22

10 Redraw each of the diagrams shown in figure 1.23 to show how **both** a voltmeter is connected to measure the voltage across component *R* **and** an ammeter is connected to measure the current through component *S*.

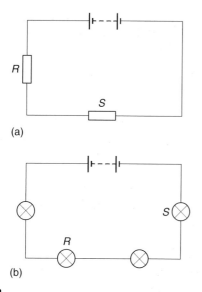

Figure 1.23

11 A current of 0.012 A passes through a 1 kΩ resistor. Calculate the potential difference across the resistor.

12 An electric toaster is connected to the 230 V mains and switched on. A current of 4.5 A passes through the toaster element. Calculate the resistance of the element.

13 The motor of a toy car is connected to a 4.5 V battery. The motor of the car has a resistance of 18 Ω. Calculate the current passing through the motor.

14 Two resistors, of value 1 kΩ and 3 kΩ, are connected in series with a 9 V battery. What is the potential difference across the 3 kΩ resistor?

2 Electrical energy

At the end of this chapter you should be able to:

1 State that when there is an electrical current in a component, there is an energy transformation.

2 State that the electrical energy transformed each second $= VI$.

3 State the relationship between energy and power.

4 Carry out calculations using $P = IV$ and $E = Pt$.

5 Explain the equivalence between VI, I^2R and V^2/R.

6 Carry out calculations involving the relationships between power, current, voltage and resistance.

7 State that in a lamp electrical energy is transformed into heat and light.

8 State that the energy transformation in an electrical heater occurs in the resistance wire.

9 Explain in terms of current the expressions d.c. and a.c.

10 State that the frequency of the mains supply is 50 Hz.

11 State that the quoted value of an alternating voltage is less than its peak value.

12 State that a d.c. supply and an a.c. supply of the same quoted value will supply the same power to a given resistor.

Power

When an electric current passes through a wire, the electrons making up the current collide with the atoms of the wire. These collisions make the atoms vibrate more and this results in the wire becoming hotter, i.e. electrical energy has been changed into heat in the wire. The amount of heat produced depends on the value of the current and the value of the resistance of the wire. Heating elements for electric fires and kettles change electrical energy into heat energy in the resistance wire making up the element.

A lamp transfers electrical energy into heat and light energies in a resistance wire called the filament. The energy transferred in one second is known as the power rating of the lamp.

Power is the energy transferred in one second.

$$\text{Power} = \frac{\text{energy transferred}}{\text{time taken}}$$

$$P = \frac{E}{t}$$

1 watt = 1 joule per second (1 W = 1 J/s)

Example

A hairdryer uses 86.4 kJ of energy in a time of 1.5 minutes. What is the power rating of the hairdryer?

Solution

$$P = \frac{E}{t} = \frac{86.4 \times 10^3}{1.5 \times 60} = 960 \text{ W}$$

Power, current and voltage

Three different lamps of known power ratings were connected to an electrical supply. The voltage (p.d.) across and the current through the lamps were recorded. The readings are shown in the table below.

Power rating of lamp	Voltage	Current
24 W	12 V	2 A
36 W	12 V	3 A
48 W	12 V	4 A

From the table we can conclude:

Power = current × voltage

$P = IV$

but from Ohm's law $V = IR$

$P = I(IR)$

$P = I^2R$

Alternatively $I = \dfrac{V}{R}$

$P = \left(\dfrac{V}{R}\right)V$

$P = \dfrac{V^2}{R}$

The equations $P = IV$, $P = I^2R$ and $P = \dfrac{V^2}{R}$ can be used to find the power rating of appliances.

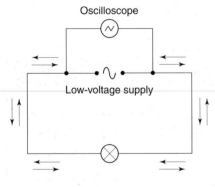

Figure 2.1 ▲
Electrons move in only one direction. This is known as direct current (d.c.)

Electron movement is to and fro. This is known as alternating current (a.c.)

Figure 2.2 ▲
Electron movement is to and fro. This is known as alternating current (a.c.)

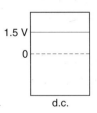

Figure 2.3 ▲
Oscilloscope trace from figure 2.1

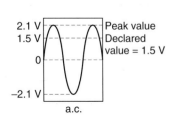

Figure 2.4 ▲
Oscilloscope trace from figure 2.2

Example

The element of an electric kettle, when operating, has a resistance of 23 Ω. The current in the element is 10 A. Calculate the power rating of the kettle.

Solution

$$P = I^2R = 10^2 \times 23 = 2300 \text{ W}$$

Example

An electric toaster has a power rating of 1050 W. Calculate the resistance of the element when it is operating from the 230 V mains supply.

Solution

$$P = \frac{V^2}{R}$$

$$1050 = \frac{230^2}{R}$$

$$R = \frac{230^2}{1050} = 50.4 \text{ Ω}$$

Direct and alternating current

Figure 2.1 shows a battery connected to a lamp. Figure 2.2 shows a mains-operated low-voltage power supply connected to an identical lamp. The lamps are equally bright.

In figure 2.1 electrons (negative charges) move from the negative terminal through the lamp and wires to the positive terminal of the battery. This means that the electrons move in only one direction – this is known as direct current or d.c.

In figure 2.2 electrons move in one direction, then in the other direction and back again, i.e. the electrons move to and fro. This alternating movement of the electrons is known as alternating current or a.c. The to and fro movement of the electrons is very frequent and occurs 50 times every second and so the frequency of mains electricity is 50 hertz (50 Hz).

The oscilloscope traces from figures 2.1 and 2.2 are shown in figures 2.3 and 2.4. The d.c. trace has a constant value of 1.5 V while the a.c. trace alternates from a maximum or peak value of +2.1V to –2.1 V.

Since both lamps are equally bright, they must be receiving and giving out the same amount of energy every second, i.e. they have the same power. The a.c. supply of 2.1 V peak is having the same effect as the d.c. supply of 1.5 V and so the a.c. supply has an effective or quoted value of 1.5 V.

The quoted (effective) a.c. value is always smaller than the peak a.c. voltage.

The alternating voltage of mains electricity in the UK is 230 volts – this is the quoted value of the voltage. The peak value is approximately 322 V.

Physics facts and key equations for electrical energy

- Power = $\dfrac{\text{energy}}{\text{time}}$, i.e. $P = \dfrac{E}{t}$.

- Power = current × voltage, i.e. $P = IV$.

- Power = current^2 × resistance, i.e. $P = I^2R$.

- Power = $\dfrac{\text{voltage}^2}{\text{resistance}}$, i.e. $P = \dfrac{V^2}{R}$.

- Mains supply has a frequency of 50 Hz.

- The declared value of mains voltage is quoted as 230 V.

- The declared value of an alternating voltage is less than the peak value.

End-of-Chapter Questions

1 The element of an electric iron is connected to the 230 V mains supply. The current in the element is 5.5 A.

   (a) Calculate the power rating of the iron.

   (b) How much electrical energy does the iron use in one second?

2 A lamp is plugged into a 12 V supply and switched on. The lamp has a resistance of 3 Ω when lit.

   (a) Write down the energy change which takes place.

   (b) Calculate the current through the lamp.

   (c) Find the power rating of the lamp.

3 The element of an electric oven has a resistance of 17.6 Ω when operating from the 230 V mains supply. Calculate the power rating of the oven.

4 A spotlight is rated at 12 V, 50 W. Calculate the resistance of the spotlight when lit.

5 A current of 0.6 A passes through an electric motor. The motor has a power rating of 138 W. Calculate the resistance of the motor when it is operating.

6 An alternating supply has a quoted value of 230 V. Give a possible value for the peak value for this supply.

7 What is the frequency of the 230 V mains supply?

3 Electromagnetism

At the end of this chapter you should be able to:

1 State that a magnetic field exists around a current-carrying wire.

2 Identify circumstances in which a voltage will be induced in a conductor.

3 State the factors which affect the size of the induced voltage, i.e. field strength, number of turns on a coil, relative movement.

4 State that transformers are used to change the magnitude of an alternating voltage.

5 Carry out calculations involving input and output voltages, turns ratio and primary and secondary currents for an ideal transformer.

Magnetic fields and electromagnetism

Permanent magnets

A magnet is able to exert a force on certain materials. The region surrounding the magnet is called a magnetic force field or simply a magnetic field. A permanent magnet, as its name implies, has a magnetic field surrounding it which cannot be switched off. The opposite ends, or poles, of a magnet are called north and south (a north pole means a north-seeking pole, i.e. it always wants to point north). The shape of the magnetic field surrounding a magnet can be shown by scattering iron filings on a piece of paper placed on top of it. The direction of the magnetic field can be found using a compass (see figure 3.1).

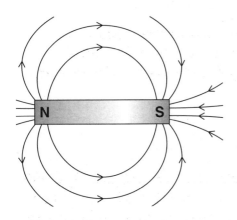

Figure 3.1 ▲
Magnetic field lines surrounding a permanent magnet

When two permanent magnets are placed close together, their magnetic fields produce forces which cause:

- A north pole to repel a north pole.
- A south pole to repel a south pole.
- A north pole to attract a south pole.

In other words, like poles repel and unlike poles attract.

Some metals such as iron and steel are attracted to a magnet.

Electromagnetism and electromagnets

Figure 3.2 shows a long straight wire passing vertically through a piece of card. A magnetic field surrounds the wire when it carries an

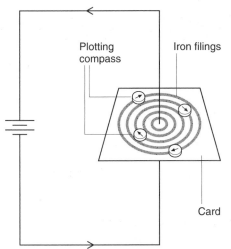

Plotting compass

Iron filings

Card

Figure 3.2 ▲
Magnetic field surrounding a current-carrying wire

Figure 3.3 ▶
An electromagnet

electrical current. Increasing the current through the wire increases the strength of magnetic field surrounding the wire. Reversing the direction of the current through the wire reverses the direction of magnetic field around the wire. When an electric current passes through a wire which is coiled around an iron core, the core becomes magnetised and an electromagnet is produced, as shown in figure 3.3. However, the electromagnet has little strength without the iron core. The iron core is able to concentrate the magnetic field within itself, so giving a stronger magnetic effect.

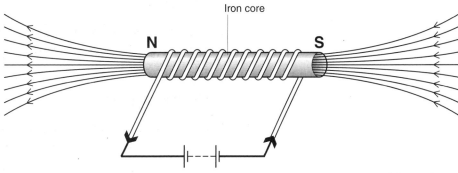

Iron core

N   S

The magnetic field of an electromagnet can be made stronger by:

- Increasing the current through the coils of wire.
- Increasing the number of turns of wire on the core.

When an a.c. supply is used, the electromagnet still produces a magnetic field – but one which alternates each time the current changes direction. There is no magnetic field when the electric current is switched off.

Generating electricity

Figure 3.4 shows a coil of wire placed between two magnets. The ends of the coil are connected to a voltmeter. When the coil is moved up or down through the magnetic field, a voltage (p.d.) is produced across the ends of the coil. The size of this voltage (p.d.) is dependent on:

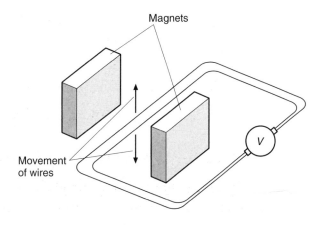

Magnets

Movement of wires

Figure 3.4 ▶
Moving coils of wire through a magnetic field to generate a voltage (p.d.)

- The number of turns of wire on the coil – the greater the number of turns, the greater the voltage produced.
- The strength of the magnetic field – the stronger the magnetic field, the greater the voltage produced.
- The speed of movement – the faster the coil is moved up or down through the magnetic field, the greater the voltage produced.

Transmitting electrical energy

Figure 3.5 shows how electrical power, generated at a power station, is distributed through the national grid transmission system to our homes.

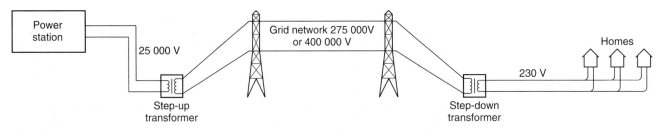

Figure 3.5 ▲
National Grid transmission system

The generator at the power station produces an output voltage of 25 000 volts. For efficient transmission over long distances this voltage is increased by a (step-up) transformer, to 275 000 volts or 400 000 volts for the national grid system. At the end of the transmission lines a (step-down) transformer reduces the voltage for distribution to consumers.

The transformer

A transformer consists of two separate coils of wire wound on the same iron core. The circuit symbol for a transformer is shown in figure 3.6. The straight line between the coils represents the iron core.

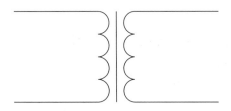

Figure 3.6 ▲
Circuit symbol for a transformer

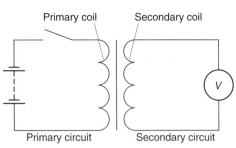

Figure 3.7 ▲
A transformer circuit

Figure 3.7 shows a transformer connected to a switch and a battery. When the switch is closed, a current passes through the primary coil. This results in a magnetic field in the primary coil which rapidly builds up through both sets of coils, since they are joined by the iron core. This gives a changing magnetic field in the secondary coil and so a voltage (p.d.) is produced.

When the current in the primary circuit is steady, there is no change in the magnetic field and no voltage (p.d.) is produced in the secondary circuit. When the switch is opened, the primary current is switched off and the magnetic field around both the primary and secondary coils rapidly collapses (disappears). This changing magnetic field through the secondary coil results in a voltage (p.d.) being produced but in the opposite direction.

Removing the iron core through the coils decreases the effect since the iron core concentrates the magnetic field through the coils.

With a d.c. supply connected to the primary coil, a changing magnetic field can only be obtained by opening and closing the switch. A more practical way of obtaining a changing magnetic field is to connect the primary coil to an a.c. supply. Now the current is always changing in size and direction. This means that the magnetic field is continually changing and so a voltage is continually produced. Transformers therefore only work on a.c.

A transformer is used to investigate the relationship between the alternating voltages at the primary and secondary coils and the number of turns on the primary and secondary coils.

Primary turns (N_p)	Secondary turns (N_s)	N_s/N_p	Primary voltage (V_p)	Secondary voltage (V_s)	V_s/V_p
125	500	4	2 V	8 V	4
500	125	0.25	2 V	0.5 V	0.25
125	625	5	2 V	10 V	5
500	500	1	2 V	2 V	1
625	125	0.2	2 V	0.4 V	0.2

From this table we can conclude that:

$$\frac{N_s}{N_p} = \frac{V_s}{V_p}$$

- In a step-up transformer: secondary turns are more than the primary turns, i.e. $N_s > N_p$, thus $V_s > V_p$.
- In a step-down transformer: secondary turns are less than the primary turns, i.e. $N_s < N_p$, thus $V_s < V_p$.

Since the energy losses in transformers are normally very small (about 5% to heat, sound and leakage of the magnetic field) it is convenient to consider the transformer as being 100% efficient, i.e. an ideal transformer.

For an **ideal** transformer all the power at the primary is transferred to the secondary. Therefore:

input power = output power

$$I_p \times V_p = I_s \times V_s$$

i.e., $$\frac{V_s}{V_p} = \frac{I_p}{I_s}$$

Hence:

$$\frac{V_s}{V_p} = \frac{N_s}{N_p} = \frac{I_p}{I_s}$$

Example

An ideal transformer steps down the 230 V mains supply to 5 V. The secondary coil of the transformer has 400 turns. Calculate the number of turns on the primary coil.

Solution

$$\frac{N_p}{N_s} = \frac{V_p}{V_s}$$

$$\frac{N_p}{400} = \frac{230}{5}$$

$$N_p = 46 \times 400 = 18,400 \text{ turns}$$

Example

An ideal transformer has 4000 primary turns and 100 secondary turns. The current in the primary coil is 30 mA. Calculate the current in the secondary coil.

Solution

$$\frac{N_s}{N_p} = \frac{I_p}{I_s}$$

$$\frac{100}{4000} = \frac{30 \times 10^{-3}}{I_s}$$

$$\frac{1}{40} = \frac{30 \times 10^{-3}}{I_s}$$

$$I_s = 30 \times 10^{-3} \times 40$$

$$I_s = 1.2 \text{ A}$$

Physics facts and key equations for electromagnetism

- A magnetic field surrounds a current-carrying wire and is controlled by the current in the wire.

- A voltage can be produced in a coil when the magnetic field near the coil changes.

- The size of the voltage produced can be increased by increasing the strength of the magnetic field, increasing the number of turns on the coil and increasing the speed of movement of the coils through the magnetic field.

- A transformer consists of two separate coils of wire wound on an iron core.

continued ➢

- Transformers are used to change the size of an a.c. voltage.

- For a transformer, $\dfrac{V_p}{V_s} = \dfrac{N_p}{N_s}$.

- For a 100% efficient transformer,
$$\dfrac{V_p}{V_s} = \dfrac{N_p}{N_s} = \dfrac{I_s}{I_p}.$$

- Electrical power is distributed by the national grid system – the output voltage of 25,000 V from the power station is stepped-up by a transformer to 400,000 V for efficient transmission along the transmission lines. At the other end of the transmission lines a step-down transformer reduces the voltage to 230 V for use in our homes.

End-of-Chapter Questions

1 The diagram shown in figure 3.8 shows some coils of wire connected to a voltmeter. When the coils are moved in the direction indicated a reading is obtained on the voltmeter. What change(s) could be made to the apparatus to increase the reading on the voltmeter?

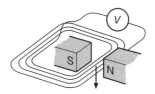

Figure 3.8

2 A 10 V a.c. supply is connected to the 40-turn primary coil of a transformer. The voltage obtained from the secondary coil is 120 V. Calculate the number of turns on the secondary coil.

3 A 120 V a.c. supply is connected to the 8000-turn primary coil of an ideal transformer. A lamp is connected to the 400-turn secondary coil of the transformer. Calculate the potential difference across the lamp.

4 The output voltage from the 750-turn secondary coil of a transformer is 5 V. The primary coil of the transformer has 1500 turns. Calculate the input voltage to the transformer.

5 An ideal transformer has a 250-turn primary coil and a 750-turn secondary coil. The current in the secondary coil is 1 A. Calculate the current in the primary coil.

6 A low-voltage heater is designed to work at 10 V and 2 A. An ideal transformer having a primary coil of 8280 turns is used with the 230 V supply.

(a) Is the supply a.c. or d.c.?
(b) Calculate the current in the primary coil.
(c) Calculate the number of turns on the secondary coil.

4 Electronic components

At the end of this chapter you should be able to:

1 Give examples of output devices and the energy conversions involved.

2 Draw and identify the symbol for an LED.

3 State that an LED will only light if connected one way round.

4 Describe by means of a diagram a circuit which will allow an LED to light.

5 Calculate the value of the series resistor for an LED and explain the need for this resistor.

6 Describe the energy transformations involved in the following devices:

microphone, thermocouple, solar cell.

7 State that the resistance of a thermistor usually decreases with increasing temperature and the resistance of an LDR decreases with increasing light intensity.

8 Carry out calculations using $V = IR$ for the thermistor and LDR.

9 Draw and identify the circuit symbol for an n-channel MOSFET.

10 Draw and identify the circuit symbol for an NPN transistor.

11 State that a transistor can be used as a switch.

12 Explain the operation of a simple transistor switching circuit.

13 Identify, from a list, devices in which amplifiers play an important part.

14 State that the output signal of an audio amplifier has the same frequency as, but a larger amplitude than, the input signal.

15 Carry out calculations involving input voltage, output voltage and voltage gain of an amplifier.

Output devices

Loudspeaker

A loudspeaker is connected to a signal generator. As the amplitude (energy of the electrical signal) from the signal generator is increased the sound from the loudspeaker gets louder. A loudspeaker changes electrical energy into sound energy.

Electric motor

An electric motor is connected to a variable d.c. supply voltage. The speed of the motor increases as the voltage (p.d.) is increased. An electric motor changes electrical energy into kinetic energy. Reversing the connections to the supply reverses the direction of rotation.

Relay

A relay is a switch operated by an electromagnet. A coil of wire, when carrying an electric current, provides the magnetic field required to close the switch contacts in the relay shown in figure 4.1. When switch S is closed a current passes through the coil surrounding the switch – the switch contacts close, completing the lower electrical circuit, thus allowing the lamp to light. When S is opened the switch contacts open and the lamp goes out. The relay changes electrical energy into the opening or closing of a switch, i.e. kinetic.

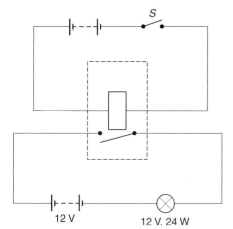

Figure 4.1 ▲
A relay circuit – closing switch S allows the lamp to light

Filament lamp

A filament lamp consists of a thin tungsten wire (filament) in a glass container. When an electric current passes through the wire, electrical energy is changed into heat and light in the filament. A lamp is connected to a variable d.c. voltage supply. Increasing the voltage (p.d.) across the lamp increases the current through it and so it gets brighter. No difference is observed when the connections from the supply to the lamp are reversed. The filament in the lamp requires a relatively large current to light properly and gets very hot in operation.

Light-emitting diode (LED)

Light-emitting diodes are made by joining two special materials together to produce a junction. When an electric current passes through the junction, electrical energy is changed into light energy. However, too large a current – or indeed too high a p.d.– will destroy the junction. To prevent this a resistor must be connected in series with the LED. Figure 4.2 shows an LED and a resistor connected to a variable d.c. voltage supply. Increasing the voltage (p.d.) across the LED increases its brightness. The LED does not light when the connections from the supply are reversed. The LED only requires a small current to light and does not get hot in operation. LEDs are available in red, green, yellow, blue and white colours.

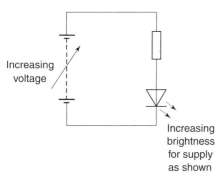

Figure 4.2 ▲
A variable voltage supply connected to a resistor and an LED

Example
The maximum voltage allowed across an LED is 1.8 V and the current through it must not exceed 10 mA. The LED is connected

to a 6.0 V d.c. supply. Calculate the value of the resistor, R, connected, in series, with the LED.

Solution

Figure 4.3 ▶
Circuit diagram

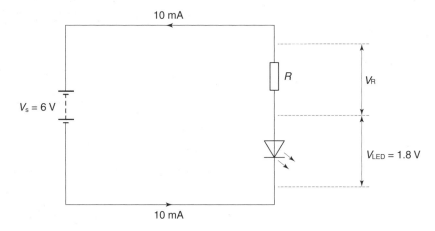

Since the LED and resistor are connected in series then $V_s = V_{LED} + V_R$ and the current through both components is the same.

Therefore $V_R = V_s - V_{LED} = 6.0 - 1.8 = 4.2$ V

$$V_R = IR$$

$$4.2 = 10 \times 10^{-3} \times R$$

$$R = \frac{4.2}{10 \times 10^{-3}} = 420 \ \Omega$$

Input devices

Microphone

A microphone is connected to an oscilloscope. As louder notes are played into the microphone, the amplitude of the trace displayed on the oscilloscope increases. A microphone changes sound energy into electrical energy. The louder the sound, the greater the electrical energy produced.

Thermocouple

A thermocouple is composed of two different types of wire joined together. A thermocouple is connected to a voltmeter as shown in figure 4.4. When the junction of the thermocouple is placed in a Bunsen flame, the voltmeter reading increases. A thermocouple changes heat energy into electrical energy. The higher the temperature of the junction, the greater the electrical energy produced.

Increasing temperature

Increasing voltmeter reading

Figure 4.4 ▲
A thermocouple connected to a voltmeter

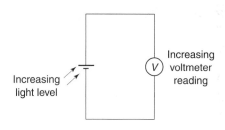

Figure 4.5 ▲
A solar cell connected to a voltmeter

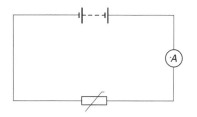

Figure 4.6 ▲
A thermistor circuit

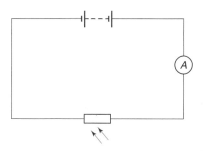

Figure 4.7 ▲
An LDR circuit

Figure 4.8 ▶
Example of a thermistor

Solar cell

A solar cell is connected to a voltmeter as shown in figure 4.5. When the solar cell is exposed to more light, the voltmeter reading increases. A solar cell changes light energy into electrical energy. The brighter the light shining on the solar cell, the greater the electrical energy produced.

Thermistor

When the thermistor shown in figure 4.6 is heated, the ammeter reading increases – therefore the resistance of the thermistor must be decreasing.

The resistance of most thermistors usually decreases with increasing temperature.

temperature ↑ – resistance of thermistor ↓

Light-dependent resistor (LDR)

When the light-dependent resistor, shown in figure 4.7, is exposed to more light, the ammeter reading increases – therefore the resistance of the LDR must be decreasing.

As the light gets brighter (light intensity increases) the resistance of the LDR decreases.

light intensity ↑ – resistance of LDR ↓

Example
A thermistor, 1 kΩ resistor and an ammeter are connected in series to a 6 V supply as shown in figure 4.8. When the thermistor is at a temperature of 20°C the reading on the ammeter is 2 mA.

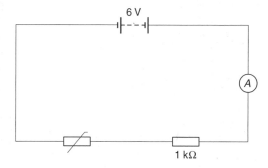

(a) Calculate the voltage across the 1 kΩ resistor at 20°C.

(b) Find the resistance of the thermistor at 20°C.

Solution

(a) $V_R = IR_R = 2 \times 10^{-3} \times 1 \times 10^3 = 2$ V

(b) $V_s = V_R + V_{\text{thermistor}}$
$6 = 2 + V_{\text{thermistor}}$

$$V_{thermistor} = 4 \text{ V}$$

But $V_{thermistor} = IR_{thermistor}$

$$4 = 2 \times 10^{-3} \times R_{thermistor}$$

$$R_{thermistor} = \frac{4}{2 \times 10^{-3}} = 2 \text{ k}\Omega$$

Example

A student uses the apparatus shown in figure 4.9 to measure the resistance of an LDR. When the LDR is illuminated the reading on the voltmeter is 2 V and the reading on the ammeter is 8 mA. Calculate the resistance of the LDR at this light level.

Solution

$$V_{LDR} = IR_{LDR}$$

$$2 = 8 \times 10^{-3} \times R_{LDR}$$

$$R_{LDR} = 250 \text{ }\Omega$$

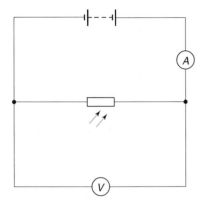

Figure 4.9 ▲
Example of an LDR

Transistors

1   A MOSFET (metal oxide semiconductor field effect transistor) – this has three terminals, called the gate, the source and the drain. The symbol for an n-channel MOSFET is shown in figure 4.10.

2   NPN transistor – this has three terminals, called the base, the emitter and the collector. The symbol for a NPN transistor is shown in figure 4.11.

The MOSFET and the NPN transistor are both types of electronic switch with no moving parts.

The switching of an NPN transistor (MOSFETs operate in a similar way) is controlled by the voltage (p.d.) applied to the emitter-base. The transistor is off (non-conducting) when the emitter-base voltage (p.d.) is below a certain value – about 0.7 V – i.e., the electronic switch is open. However, the transistor is on (conducting) when the emitter-base voltage (p.d.) is equal to or above this certain value (≥0.7 V), i.e. the electronic switch is closed.

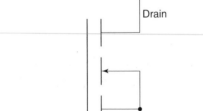

Figure 4.10 ▲
MOSFET symbol

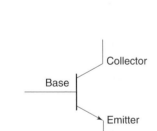

Figure 4.11 ▲
NPN transistor symbol

A temperature-controlled circuit

Figures 4.12 and 4.13 show two temperature controlled circuits. The variable resistor, in each circuit, is adjusted until at room temperature the LED is just off.

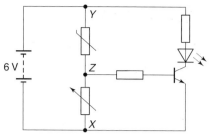

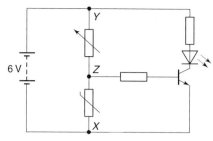

Figure 4.12 ▲

A temperature-controlled circuit

LED is off

Heat thermistor

$R_{thermistor} \downarrow$

$V_{thermistor} \downarrow$ so $V_{zy} \downarrow$

$V_{xz} \uparrow$

Transistor switches on

LED lights

Figure 4.13 ▲

Alternative temperature-controlled circuit

LED is off

Cool thermistor

$R_{thermistor} \uparrow$

$V_{xz} \uparrow$

Transistor switches on

LED lights

Note. A variable resistor is used in this type of circuit instead of a fixed resistor. The variable resistor allows the circuit to be adjusted to different conditions (temperature in this case) before the output device comes on (or goes off).

A light-controlled circuit

Figures 4.14 and 4.15 show two light-controlled circuits. The variable resistor is adjusted, in each circuit, until at normal light level the LED is just off.

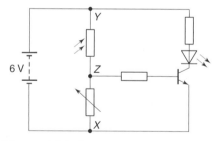

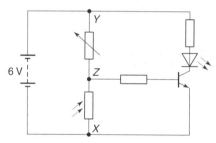

Figure 4.14 ▲

A light-controlled circuit

LED is off

Shine more light on LDR

$R_{LDR} \downarrow$

$V_{LDR} \downarrow$ so $V_{ZY} \downarrow$

$V_{XZ} \uparrow$

Transistor switches on

LED lights

Figure 4.15 ▲

Alternative light-controlled circuit

LED is off

Cover LDR

$R_{LDR} \uparrow$

$V_{XZ} \uparrow$

Transistor switches on

LED lights

Amplifiers

An amplifier is a device which is generally used to make electrical signals larger. Figure 4.16 shows an amplifier being used to make the electrical signal from a signal generator larger. The traces produced by both input and output signals from the amplifier are displayed on the screens of identical oscilloscopes as shown in figure 4.17. The amplitude of the trace of the output signal from the amplifier is larger than the amplitude of the trace of the input signal – the extra energy comes from the electrical supply to the amplifier. The frequency of the output from the amplifier is the same as the frequency of the input signal. Amplifiers are to be found in most audio devices, e.g. radios, televisions, hi-fi's, intercoms and loudhailers.

Figure 4.16 ▶
An amplifier circuit

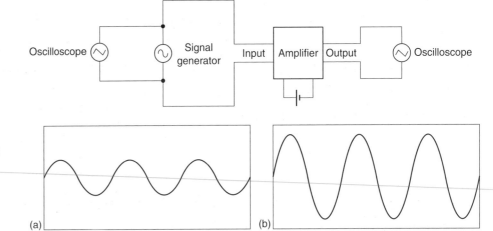

Figure 4.17 ▶
(a) Input voltage to amplifier (b) Output voltage from amplifier

Voltage gain

An amplifier has an input voltage of 0.5 V and an output voltage of 5.0 V – the output voltage is 10 times larger than the input voltage, i.e. the voltage gain is 10:

$$\text{Voltage gain} = \frac{\text{output voltage}}{\text{input voltage}}$$

Note that voltage gain does not have a unit.

Example

The voltage gain of an amplifier is 200. Calculate the voltage at the output of the amplifier when the voltage at its input is 20 mV.

Solution

$$\text{Voltage gain} = \frac{\text{output voltage}}{\text{input voltage}}$$

$$200 = \frac{\text{output voltage}}{20 \times 10^{-3}}$$

$$\text{output voltage} = 200 \times 20 \times 10^{-3} = 4 \text{ V}$$

- An output device changes electrical energy into some other form of energy.

- There are a number of output devices: examples are loudspeaker, relay, filament lamp and light-emitting diode (LED).

- LEDs will only light up when connected to the power supply the correct way round.

- A resistor should always be connected in series with an LED – this protects the LED from damage from too high a current passing through it (or too high a voltage across it).

- Most input devices change some form of energy into electrical energy.

- There are a number of input devices: examples are microphone, thermocouple, solar cell, thermistor and light-dependent resistor (LDR).

- The resistance of a thermistor changes with temperature – for the thermistors considered in this book the resistance of a thermistor decreases as the temperature increases.

- The resistance of an LDR decreases with increasing light level.

- An NPN transistor and a MOSFET are electrically operated switches.

- A transistor is non-conducting (OFF) for voltages below a certain value (normally below 0.7 V) but conducting (ON) at voltages at or above this certain value (at or above 0.7 V).

- An amplifier is a device which makes electrical signals larger.

- Audio amplifiers are found in devices such as radios, televisions, hi-fi's, intercoms and loudhailers.

- The output signal from an audio amplifier has the same frequency as, but a larger amplitude than, the input signal.

- For an amplifier,

$$\text{voltage gain} = \frac{\text{output voltage}}{\text{input voltage}}$$

End-of-Chapter Questions

1 Name two output devices and state the energy conversion for each one.

2 A student designs a suitable circuit to light an LED. The student uses the following components: a 6 V battery, a switch, an LED and a resistor.

   (a) Draw a suitable diagram, which will allow the LED to light when the switch is closed.

   (b) The maximum voltage across the LED must not exceed 1.75 V and the maximum current must not exceed 11 mA. Calculate the value of the resistor required for the circuit.

3 State the energy conversion for: (a) a microphone, (b) a thermocouple, (c) a solar cell.

4 A thermistor is connected to a power supply. A student connects an ammeter and a voltmeter in the circuit so that she can find the resistance of the thermistor. When the temperature of the thermistor is 18°C the reading on the voltmeter is 6 V and the reading on the ammeter is 12 mA.

   (a) Calculate the resistance of the thermistor at this temperature.

   (b) The temperature rises to 20°C. The voltmeter reading remains at 6 V. Suggest a value for the ammeter reading. continued ➤

End-of-Chapter Questions continued

5   Figure 4.18 shows a circuit built by a student to detect when the element of an electric cooker is hot. The resistance of the thermistor decreases as its temperature increases.

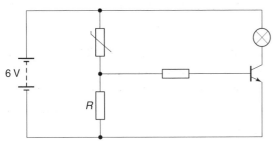

Figure 4.18 ▲

Explain what will happen as the temperature of the cooker element increases.

6   Draw the circuit symbol for: (a) an NPN transistor, (b) a MOSFET.

7   The input voltage to an audio amplifier is 10 mV and the output voltage from it is 3.0 V.

(a) Calculate the voltage gain of this amplifier.

(b) How do the frequencies of the input and output signals of an amplifier compare?

8   The voltage gain of an amplifier is 200. Calculate the voltage at the output of the amplifier when the voltage at its input is 3 mV.

Exam Questions

1   (a) Two resistors are connected in series to a 9 V supply as shown in figure E.1.

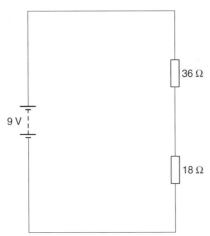

Figure E.1 ▲

(i) Calculate the current drawn from the supply.
(ii) Find the voltage across the 18 Ω resistor.

(b) An electrical appliance has three resistors connected as shown in figure E.2.

(i) Calculate the resistance between points A and B.

(ii) Calculate the power rating of the appliance when points A and B are connected to the 230 V mains supply.

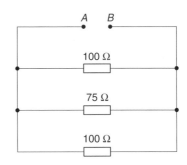

Figure E.2 ▲

2   A set of Christmas tree lights consists of 19 lamps and a resistor, all connected in series. The lamps are labelled 12 V, 3 W. The set is connected to the 230 V mains supply and switched on.

(a) Why must the 19 lamps and resistor be connected in series?

(b) Calculate the current through each of the lamps.

(c) Show that the resistance of the resistor is 8 Ω.

continued ➢

(d) How much charge flows through each lamp in 1 minute?

(e) Calculate the resistance of each lamp.

3 The element of an electric kettle has a power rating of 2116 W. The element is connected to the 230 V mains and switched on for 180 s.

(a) (i) How much electrical energy is transferred into heat in 180 s?

(ii) Where does this energy change occur?

(b) Calculate the resistance of the element when it is switched on.

4 An electrical component and a resistor are connected in series with a 10 V supply as shown in figure E.3.

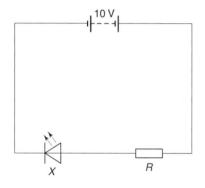

Figure E.3 ▲

(a) Name component X.

(b) Explain why resistor R is necessary in the circuit.

(c) The voltage across component X is 1.8 V and the current in the circuit 11 mA. Calculate the value of resistor R.

5 An ideal transformer has a 2000-turn primary coil and a 200-turn secondary coil. A 12 V, 24 W lamp is connected to the secondary coil of the transformer. The lamp is working at its rated voltage.

(a) Calculate the input voltage to the primary coil of the transformer.

(b) Find the current in the lamp.

(c) Find the current in the primary coil of the transformer.

Mechanics and Heat

5 Kinematics

At the end of this chapter you should be able to:

1 Describe how to measure an average speed.

2 Carry out calculations involving the relationship between distance, time and average speed.

3 Describe how to measure instantaneous speeds.

4 Identify situations where average and instantaneous speed are different.

5 Describe what is meant by vector and scalar quantities.

6 State the difference between distance and displacement.

7 State the difference between speed and velocity.

8 Explain the terms speed, velocity and acceleration.

9 State that acceleration is change in velocity per unit time.

10 Draw velocity–time graphs of more than one constant acceleration.

11 Describe the motions represented by a velocity–time graph.

12 Calculate displacement and acceleration, from velocity–time graphs, for more than one constant acceleration.

Speed

Speed is the distance travelled by a vehicle or a person in one second. Your speed will vary throughout a journey and we will talk about average speed:

$$\text{average speed} = \frac{\text{distance travelled}}{\text{time taken}}$$

In symbols, $v = \dfrac{d}{t}$

Average speed is the constant speed needed to cover the distance travelled in the time. Speed and average speed have the same units of m/s.

Measuring average speed

Using a stopwatch, trolley and measuring tape

The distance is measured between two points a few metres apart on the ground. The time for the journey is measured by starting a stopwatch when the trolley reaches the first point (*x*) and stopping the watch when it reaches the second point (*y*) (figure 5.1).

Figure 5.1 ▶
Measuring average speed

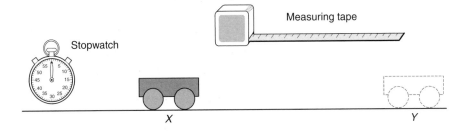

Instantaneous speed

This is the measure of speed at one instant of time – for example, just as a runner crosses a finishing line or a car passes on the bend of a racing track. If we could measure very small time intervals such as thousandths of a second then we could measure the speed of any object just at that moment in time.

Instantaneous speed is the speed of an object at a particular time (instant).

$$\text{average speed} = \frac{\text{total distance travelled}}{\text{total time taken}} = \frac{d}{t}$$

When the time taken (*t*) is very small, the closer the average speed is to the instantaneous (actual) speed.

Example

A car moves a distance of 150 m in a time of 7.5 s. Calculate the car's average speed.

Solution

$$\text{average speed} = \frac{150}{7.5} = 20 \text{ m/s}$$

This time interval is too large to give a reasonable estimate of the instantaneous speed. If we use a much smaller distance and use a smaller time interval we can obtain a measure of the instantaneous speed. This is shown in figure 5.2.

Figure 5.2 ▶
Measuring instantaneous speed

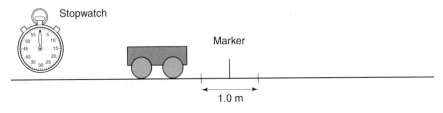

$$\text{distance} = 1.0 \text{ m}$$

$$\text{time to travel} = 0.05 \text{ s}$$

$$\text{average speed} = \frac{1.0}{0.05} = 20 \text{ m/s}$$

This is closer to the instantaneous speed as the time interval involved is very small.

This means that instantaneous speed = average speed of an object, provided the time used is very small. However, this small time interval is very difficult to measure since human reaction time is involved.

Measuring instantaneous speed using a computer

The computer uses an internal clock, which allows very small time intervals to be measured. This allows the calculation of instantaneous speed if a distance is measured. The computer needs to be connected to a piece of equipment called a light gate.

The computer starts timing when a light beam in the light gate is cut by a card, and stops when the light beam is restored. The time taken for the card to pass through the beam is recorded in the computer (figure 5.3).

Figure 5.3 ▶

Measuring instantaneous speed using a computer

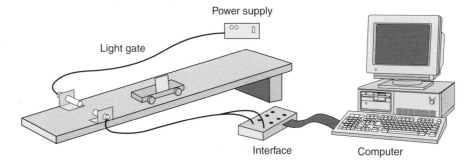

The computer has been told the length of the card and calculates the speed of the vehicle using:

$$\text{speed} = \frac{\text{length of card}}{\text{time on computer}}$$

Typical results might be:

Length of card = 5 cm = 0.05 m

Time measured by computer = 0.02s

$$\text{Speed} = \frac{0.05}{0.02}$$

$$= 2.5 \text{ m/s}$$

The police measure both average speed and instantaneous speed.

Average speed can be found from the latest cameras, which time between two marks on the ground. Since the distance between the marks is known, the average speed can be calculated (figure 5.3).

In a similar way the single speed camera can be used. This uses a sensor to detect when a car is possibly speeding. The camera takes two pictures which are 0.5 seconds apart. The road surface has markings which are 2 m apart, allowing the speed to be found (figure 5.4).

Figure 5.4 ▶
Speed camera

The speed gun which is used to check speeds measures an instantaneous speed. High-frequency waves are sent out from the gun and return from a moving object at a different frequency. The difference in frequency can be used to calculate the speed (figure 5.5).

Figure 5.5 ▶
Speed gun

Vectors and scalars

When you travel in a car or an aircraft, the driver or pilot can tell you the speed at which the vehicle is travelling. This tells you nothing about the direction of travel. If you specify the direction then you have a vector.

A scalar has only magnitude. Examples are mass, distance, speed. These quantities can be added in the normal way using arithmetic: 5 kg of potatoes and 3 kg of potatoes give us 8 kg of potatoes! This tells us all we wish to know about potatoes.

Vectors have magnitude and direction. Vectors must be added using the ideas of a scale diagram since vectors are represented by straight lines drawn to scale.

Displacement is distance with a direction. Examples are 2 km due west; 500 m at a bearing of 023°. It is the vector showing the distance and direction from your starting point to your finishing point in a straight line.

Example

Lesley walks 7 km south and then 3 km north. What is the displacement from the starting point?

Solution

The two lines are shown in figure 5.6.

- The lines are shown with an arrow to indicate direction and the tail of one vector should join at the head of the other.
- The vectors should be drawn to scale and the scale stated.
- A line from the tail of the first vector to the head of the second represents the resultant vector.
- The final displacement is 4 km.

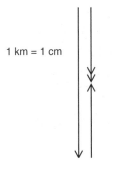

1 km = 1 cm

Figure 5.6 ▲
Vectors and displacement

Similarly velocity is a vector since it is speed with a direction:

$$\text{velocity} = \frac{\text{displacement}}{\text{time}}$$

An aircraft's velocity is 100 m/s at 069°.

Acceleration

Acceleration (a) is the change in velocity of an object in one second:

$$\text{acceleration} = \frac{\text{change in velocity}}{\text{time for change}} = \frac{\text{final velocity} - \text{initial velocity}}{\text{time for change}}$$

$$a = \frac{v - u}{t}$$

where v = final velocity, u = initial velocity and t = time for change in velocity to occur.

Acceleration is usually calculated in metres per second per second. We write this as m/s^2.

A car has an acceleration of 5 m/s^2. This means that the speed of the car increases by 5 m/s in every second. If the car starts from rest then:

after 1 s, velocity = 5 m/s

after 2 s, velocity = 10 m/s

after 3 s, velocity = 15 m/s

after 4 s, velocity = 20 m/s

Example

A cheetah starting from rest accelerates uniformly and can reach a velocity of 24 m/s in 3 seconds. What is the acceleration?

Solution

$$u = 0; v = 24 \text{ m/s}; a = ?; t = 3 \text{ s}$$

$$\text{Acceleration} = \frac{v - u}{t} = \frac{24 - 0}{3} = \frac{24}{3} = 8 \text{ m/s}^2$$

When an object is slowing down it will have a negative value for its acceleration – this is called a deceleration.

Example

A student on a scooter slows down uniformly from 6 m/s to 2m/s in 4 seconds. Calculate her deceleration.

Solution

$$u = 6 \text{ m/s}; v = 2 \text{ m/s}; a = ?; t = 4 \text{ s}$$

$$\text{Acceleration} = \frac{v - u}{t} = \frac{2 - 6}{4} = \frac{-4}{4} = -1 \text{ m/s}^2$$

$$\text{Deceleration} = 1 \text{ m/s}^2$$

Velocity–time graphs

If we measure velocity and time we can draw graphs. These graphs allow us to see the motion of an object more clearly than just looking at a table of results. It also allows us to make calculations from the graph to give additional information.

Three types of motion are shown in figure 5.7:

(a) *Uniformly increasing velocity.* This is a straight line at an angle to the horizontal. The gradient of the graph allows us to calculate the acceleration.

(b) *Steady or constant velocity.* This is a straight line parallel to the time axis.

(c) *Uniformly decreasing velocity.* This is a straight line heading down towards the time axis. Again the gradient can be used for the acceleration.

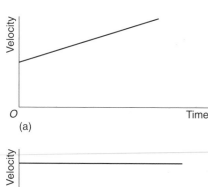
(a)

(b)

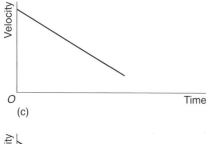
(c)

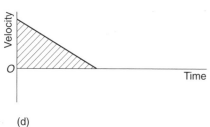
(d)

Figure 5.7 ▲
(a) Uniformly increasing velocity.
(b) Constant velocity.
(c) Uniformly decreasing velocity.

You should note the following:

- The area under any part of a velocity–time graph is the displacement of the object. Normally the shape of the graph will consist of a rectangle and/or triangles (Figure 5.7d).
- If the object is travelling in one direction in a straight line then distance travelled = displacement.
- To calculate the average speed of an object when more than one type of motion is involved, draw a speed–time graph and find the area under it, i.e. distance travelled, then:

$$\text{average speed} = \frac{\text{distance travelled}}{\text{time taken}}$$

- For an object moving only with constant acceleration or constant deceleration:

$$\text{average velocity} = \frac{\text{initial velocity} + \text{final velocity}}{2}$$

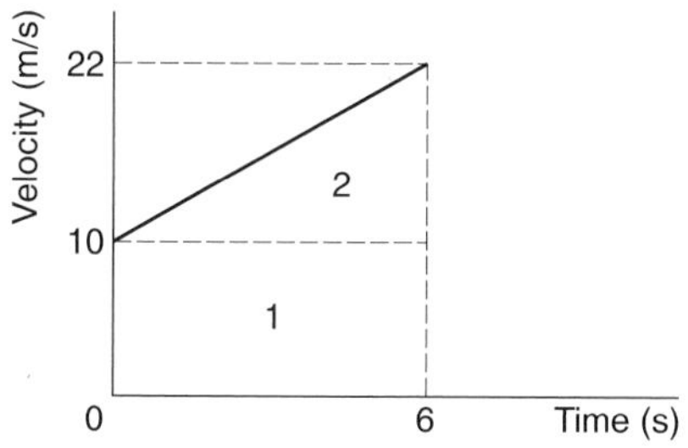

Example

The velocity–time graph for a car travelling along a road is shown figure 5.8:

(a) Calculate the acceleration of the car.
(b) Calculate the distance travelled by the car.
(c) Calculate the average speed of the car.

Solution

(a) The acceleration is given as $a = \dfrac{v - u}{t}$

where $u = 10$ m/s
$\qquad v = 22$ m/s
and $\qquad t = 6$ s

$$a = \frac{22 - 10}{6} = \frac{12}{6} = 2 \text{ m/s}^2$$

(b) Distance $\;=$ area under velocity–time graph
$\qquad\qquad\quad =$ area 1 + area 2
$$= 10 \times 6 + \frac{1}{2} \times 12 \times 6$$
$$= 60 + 36$$
$$= 96 \text{ m}$$

(c) Average speed $= \dfrac{\text{distance}}{\text{time}} = \dfrac{96}{6} = 16$ m/s

Figure 5.8 ▲
Velocity–time graph for a car travelling along a road

- average speed = $\dfrac{\text{distance.}}{\text{time}}$

  Unit of speed is m/s.

- Instantaneous speed is the same calculation as average speed but the time interval is very short.

- Vectors have magnitude and direction but scalars have only magnitude.

- Displacement and velocity are vectors since they have direction as well as magnitude but distance and speed are scalars.

- Acceleration $a = \dfrac{v - u}{t}$ and has units of m/s^2.

- Displacement is the area under a velocity–time graph.

- Acceleration is the gradient of a velocity–time graph.

End-of-Chapter Questions

1 A car travels a distance of 25 m in 4.25 s. Calculate the average speed.

2 In a police speed check, photographs are taken 0.5 s apart. The marks are 2 m apart and the car crosses 4 marks. If the speed limit for this section of the road is 13.3 m/s, calculate if the car was speeding. Is this an average or instantaneous speed?

3 During a charity walk, some pupils walk round a circular track of circumference 30 m. One complete circuit takes 15 s. Calculate the average speed and the average velocity.

4 Explain the difference between vectors and scalars and give an example.

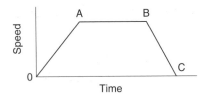

Figure 5.9 ▲

5 The graph shown in Figure 5.9 shows the motion of a car during a journey. Describe the motion of the car during the different stages.

6 A car travels at 10 m/s and increases its speed to 19 m/s in 3 s. Calculate the acceleration.

7 A lorry brakes from 20 m/s to 8 m/s in 2.4 s. Calculate the deceleration.

8 A velocity-time graph for a car journey is shown in Figure 5.10. Calculate the displacement of the car during the journey.

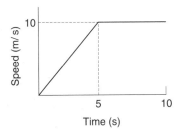

Figure 5.10 ▲

9 An electric car travels slowly along a straight road and the velocity-time graph is shown in Figure 5.11. Calculate the acceleration.

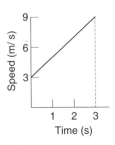

Figure 5.11 ▲

6 Forces

After reading this chapter you should be able to:

1 Describe the effects of forces in terms of their ability to change the shape, speed and direction of travel of an object.

2 Describe the use of a Newton balance to measure force.

3 State that weight is a force and is the Earth's pull on the object.

4 Distinguish between mass and weight.

5 State that weight per unit mass is called the gravitational field strength.

6 Carry out calculations where the value of the gravitational field strength is not 10 N/kg.

7 State that the force of friction can oppose the motion of a body.

8 Describe and explain situations in which attempts are made to increase or decrease the force of friction.

9 State that force is a vector quantity.

10 State that equal forces acting in opposite directions on an object are called balanced forces and are equivalent to no force at all.

11 Explain the movement of objects in terms of Newton's first law.

12 Describe the qualitative effects of change of mass or of force on the acceleration of an object.

13 Define the newton.

14 Use free body diagrams to analyse the forces on an object.

15 State what is meant by the resultant of a number of forces.

16 Use a scale diagram to find the resultant of two forces acting at right angles to each other.

17 Carry out calculations using the relationship between a, F and m and involving more than one force but in one dimension only.

18 Explain the equivalence of acceleration due to gravity and gravitational field strength.

19 Explain the curved path of a projectile in terms of the force of gravity.

20 Explain how projectile motion can be treated as two independent motions.

21 Solve numerical problems on projectiles.

Forces and motion

The effects of forces

When an object is pushed or pulled, a force is exerted on it.

Forces can have three effects on an object:

- A force can change the speed of a moving object.
- A force can change the direction of a moving object.
- A force can change the shape of (deform) an object.

These effects will depend on the size of the force applied to the object.

Measuring force

Springs can be used to measure force:

1 A spring stretches evenly – each time a mass is added to the carrier the spring stretches by the same amount.

2 A spring returns to its original length when the force is removed. If the force is too great the spring will not return to its original length when the force is removed.

We measure forces using a Newton balance. The unit of force is the **newton** (N) (figure 6.1).

Figure 6.1 ▲
The Newton balance

Frictional forces

Moving vehicles such as cars can slow down due to forces acting on them. These forces can be due to the road surface and the tyres, the brakes, or even air resistance. The force that tries to oppose motion is called the force of friction. A frictional force always acts when particles are sliding across one another and will oppose any motion. It will try to slow down an object like a car regardless of the direction in which it is moving along a road.

Car brakes

Car brakes operate by slowing down the car. When the brake pedal is pressed it causes the expander to push the brake shoes against the drum. If the brakes lock it will be very difficult to steer (figure 6.2).

Many cars are fitted with an anti-lock braking system called ABS. This uses sensors located at each wheel to detect when it is about to lock during braking. This operates by rapidly releasing and reapplying the brakes (figure 6.3).

Figure 6.2 ▶
Car braking system

Figure 6.3 ▶
Car anti-lock braking system

Reducing friction

Lubrication

The frictional force between two surfaces moving against each other can be reduced by lubricating the surfaces. This generally means that oil can be placed in between two metal surfaces. This happens in car engines and reduces wear on the engine since the metal parts are not actually meeting each other but have a thin layer of oil between them.

Streamlining

Modern cars are designed to offer as little resistance (drag) to the air as possible. This is the friction of the air on the car. The force depends on both the speed of the car and the frontal area presented to the air flow. To reduce this friction the designers try to streamline the vehicle in a variety of ways. This streamlining is measured by a number called the drag coefficient, C_d. The larger the C_d number, the greater the resistance to air flow. The calculation of C_d is difficult and involves the vehicle being placed in a wind tunnel and smoke flowing over it, as shown in figure 6.4. The more turbulence that is created the higher the value of C_d.

Figure 6.4 ▶
Drag flow

Most C_d values range from 0.3 to 0.4 since cars have similar shapes. The drag coefficient can be reduced in a number of ways:

- Reducing the front area of the car
- Using door mirrors instead of wing mirrors.
- Having a smooth round body shape.
- Using aerials made as part of the car windows.

In addition you should not carry a roof rack unless it is needed.

Many modern cars look very similar since they are all designed to this brief. Older cars often have more distinctive features but they are less fuel efficient.

Newton's first law

Sir Isaac Newton was appointed Lucasian Professor of Mathematics at Cambridge, where he worked on mathematics and astronomy. The physicist Stephen Hawking, author of *A Brief History of Time*, holds the same post today (figure 6.5).

Newton proposed three laws of motion. His first law states:

An object will remain at rest or move at constant speed in a straight line unless acted on by an unbalanced force.

Figure 6.5 ▶
(a) Sir Isaac Newton.
(b) Professor Stephen Hawking.

(a)  *(b)*

Using this law we can explain the following situations:

1 An ocean liner travelling at a steady speed through the water (figure 6.6). The resistive force offered by the water will balance out the force exerted by the engine.

Figure 6.6 ▶
Forces acting on an ocean liner

Engine force ⇐ Resistive force ⇒

2 A motorcycle moving at a constant speed along a level road and no matter how hard the driver tries to accelerate the motorcycle will not increase its speed (Figure 6.7). Again the resistive forces produced by the air frictional force will balance the force produced by the engine.

Figure 6.7 ▶
Forces acting on a motorcycle

Engine force ⇐ Air resistance ⇒

Mass and weight

Mass

Mass is the quantity of matter forming an object. This depends on the number and type of atoms making it up, so the mass of an object remains constant. If you go to different parts of the Earth or even to different planets you have the same mass since there are no changes to the number of atoms in your body.

Force of gravity

Force of gravity is the downward pull of the earth on an object:

force of gravity = pull of Earth on an object

= gravitational force on an object

= weight of an object

Weight

The weight of an object depends on:

(a) Mass.
(b) Where it is on the Earth or in the solar system.

It is a force and is measured in newtons (N).

$$\frac{\text{weight of an object}}{\text{mass of an object}} = \frac{W}{m} = \text{constant}$$

This constant is called the gravitational field strength (g).

Gravity varies depending on where you are both on the earth and in space. For instance, on Earth $g = 10$ N/kg but on the moon $g = 1.6$ N/kg and on Jupiter $g = 26.4$ N/kg. The units of g are N/kg. Thus:

$$\text{weight} = \text{mass} \times \text{gravitational field strength}$$
$$W = mg$$

Example
What is the weight of a 75 kg person on Earth?

Solution

$$W = mg$$
$$= 75 \times 10$$
$$= 750 \text{ N}$$

Acceleration due to gravity

The force of gravity causes all objects to accelerate as they fall back to earth. This is due to the earth being massive and it exerts a very strong force on these small objects. This gravitational force is the one force that we cannot switch off.

$$\text{gravitational field strength} = \frac{\text{force of gravity on an object}}{\text{mass of object}}$$

$$g = \frac{\text{weight}}{\text{mass}} = \frac{W}{m} = \text{acceleration due to gravity}$$

Forces and Supported Bodies

We know from Newton's first law that if a body either remains stationary, or moves upwards or downwards at constant speed, then the upward force is equal in size to the weight of the body but acts in the opposite direction. This is known as balanced forces.

Example 1
A stationary mass m hangs from a rope. The weight of the body is m g (N) and acts downwards, but this is counterbalanced by a force of the same size acting upwards due to the tension in the string (figure 6.8).

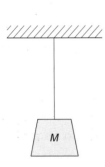

Figure 6.8 ▲
A suspended mass

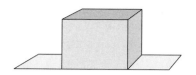

Figure 6.9 ▲
A book resting on a shelf

Example 2

A book rests on a shelf. The shelf gives an upward force equal in size to the weight, and therefore the forces are balanced (figure 6.9).

We can also see an example of balanced forces in action in a different situation:

Aircraft

The engines provide a forward force or **thrust** which accelerates the aircraft forward. However, as it moves faster the air resistance, or drag, increases until the forces (horizontally) are balanced. The aircraft would move at constant speed (Newton's first law). When in level flight at constant speed the increased **lift** will balance the weight (figure 6.10). Horizontally the thrust equals the drag force, which produces balanced forces.

Figure 6.10 ▶
Forces acting on an aircraft

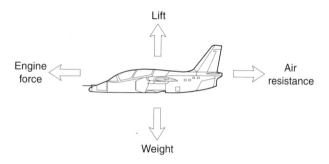

Inertia

All objects are unwilling (or reluctant) to change their motion. This means that a steady force is required to change the motion of a body.

This reluctance of a body to change its motion is called its inertial mass or mass of inertia. It depends on mass. The larger the mass, the larger the inertia and the more unwilling the body is to change its motion. It is easier to stop a lighter object than a more massive one travelling at the same speed.

Seat belts

During the sudden braking of a car, any unrestrained object will continue to move at the car's original speed. This happens due to Newton's first law. Any objects will probably collide with some part of the interior, causing damage or injury.

A seat belt applies a force in the opposite direction of motion causing rapid deceleration of the wearer. The webbing straps are designed to have a certain amount of 'give' so that the sudden force applied to the person does not cause injury (figure 6.11).

Figure 6.11 ▼
Seat belt

Adult seat belts are not suitable for young children since the young children tend to slip down in the seat. This effect is called submarining and can be prevented by using a special design for children.

Air bags produce a similar effect and the large area of the bag prevents chest injuries due to the steering wheel. This is because the larger area of the bag spreads out the force, reducing the pressure.

Whiplash injuries occur when the head of the person in an accident stays stationary but the trunk moves forward. This is shown in figure 6.12. In (a) the trunk is moved forward since the vehicle is struck from behind. The inertia of the head (b) causes it to stay in place while the trunk of the body moves forward. This causes stretching in the neck region. As the vehicle slows the head is accelerated forward. In a typical collision, at 15 m/s (about 34 mph) the collision with the vehicle will cause these injuries to occur in milliseconds.

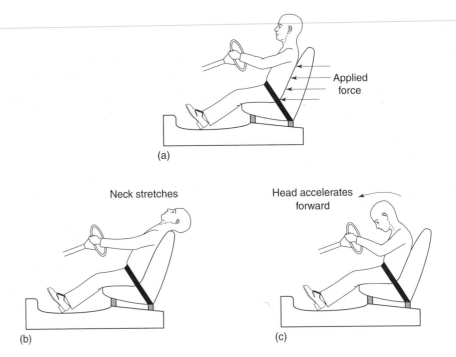

Figure 6.12 ▶
Whiplash injury

Frictional forces in a fluid

A fluid is a liquid or gas. In a fluid the frictional force on a body travelling through it increases as the speed of the body increases. If two spheres are dropped into different liquids such as oil and water the one in the water will reach the bottom first.

In both cases the spheres accelerate and then move with constant speed (terminal velocity). When the sphere is accelerating there is an unbalanced force acting on it. When the terminal velocity is reached the forces are balanced. It is reached sooner in the thick liquid than in water. This shows that the frictional force in a liquid depends on the type of liquid.

The motion of a body falling through any fluid can be divided into three parts:

1 Initially an unbalanced force acts on the body due to its weight and the body falls with a constant acceleration of 10 m/s², which is the acceleration due to gravity.

2 After a short time the frictional force begins to act and alters the motion. This force will be increasing as the speed of the body increases. There is a smaller unbalanced force ($F_{un} = W - F_r$) and the acceleration is therefore less than 10 m/s². This acceleration will continue to decrease as the frictional force increases.

3 Finally the frictional force balances the weight of the object. We now have balanced forces. The object now falls at a constant speed in a straight line. The body has reached its greatest speed. This is called its terminal speed or terminal velocity (figure 6.13).

A free-falling parachutist is a typical example of such forces in action. Figure 6.14 shows a graph of speed against time for a parachutist falling in free fall out of an aircraft and then opening the rip cord some time later. Each part of the graph is explained below.

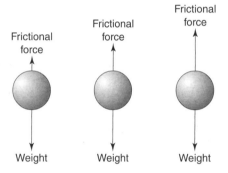

Figure 6.13 ▲
Motion of a body falling through a fluid

Figure 6.14 ▶
Motion of a parachutist

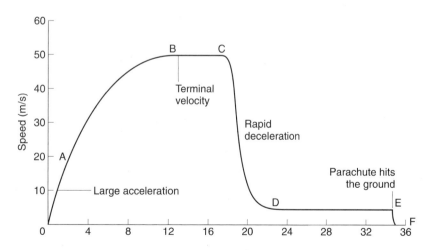

OA: constant acceleration due to gravity

AB: decreasing acceleration as frictional force acts:

unbalanced force = weight – frictional force

BC: constant speed as frictional force upwards = weight downwards

CD: non-uniform deceleration due to parachute opening and increasing frictional force

DE: constant speed as frictional force upwards = weight downwards

EF: body hits ground.

Balanced and unbalanced forces

Balanced forces acting on an object are equal in size but opposite in direction.

They cancel each other out and thus are the equivalent of zero force acting on the object (figure 6.15).

An unbalanced force acting on an object causes it to speed up or slow down.

When A and B are both same size, the same size of force is applied to each side of the vehicle. The forces are balanced. If the vehicle is at rest it will stay at rest.

If the vehicle is moving at a constant speed in a straight line it will continue at that speed in a straight line.

When the engine force A is greater than the air resistance force B, the car will accelerate to the right.

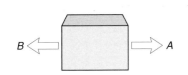

Figure 6.15 ▲
Balanced forces acting on an object

Force, mass and acceleration

If a mass m is accelerated by an unbalanced force (F) then if we double the unbalanced force we double the acceleration.

If you keep the unbalanced force constant and double the mass then you will halve the acceleration. This is Newton's second law.

When a body is acted on by a constant unbalanced force, the body moves with constant acceleration in the direction of the unbalanced force:

$$\text{unbalanced force} = \text{mass} \times \text{acceleration}$$

$$F = ma$$

In symbols in units where F is in newtons (N), m is in kilograms (kg) and a is in metres per second squared (m/s^2).

Example
An unbalanced force of 75 N acts on a mass of 15 kg. Calculate the acceleration of the mass.

Solution

$F_{un} = 75$ N, $m = 15$ kg, $a = ?$

$F_{un} = ma$

$$a = \frac{F_{un}}{m} = \frac{75}{15} = 5 \text{ m/s}^2$$

Example

A sledge is being pulled by an unbalanced force of 100 N. The sledge accelerates at 5 m/s². What is the mass of the sledge?

Solution

$F_{un} = 100$ N

$a = 5$ m/s²

$F_{un} = ma$

$100 = m \times 5$

$$m = \frac{100}{5}$$

$$= 20 \text{ kg}$$

Example

A car is being pulled by a constant force of 2000 N. The car has a mass of 1000 kg. There is a frictional force of 100 N. Calculate the acceleration of the car.

2000 N 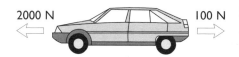 100 N

Figure 6.16 ▲

Solution

The actual force acting on the car $F_{un} = 2000 - 100 = 1900$ N (figure 6.16):

$$a = \frac{F_{un}}{m} = \frac{1900}{1000} = 1.9 \text{ m/s}^2$$

Force as a vector

When two forces act on an object the combined effect depends on the size and direction of the forces.

Force is a vector; that is, it has magnitude and direction.

The combined effect of the forces is called the resultant force. The resultant force can be found using the following technique:

- Sketch a diagram of the situation.
- Draw known lines to represent the forces and the angles.
- The lines are shown with an arrow to indicate direction.
- The tail of one vector should join the head of another vector.
- The resultant force is represented by a line drawn from the tail of the first vector to the head of the last vector.

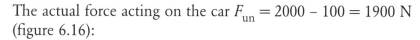

- Use a suitable scale or use Pythagoras.
- Calculate the resultant force.
- Calculate or measure the angle of the resultant force.

Example

A parachutist is falling to the ground. She has a weight of 60 N and the wind force is 15 N westwards. Calculate the resultant force on the parachutist.

Solution

Using the diagram in figure 6.17, lines are drawn to represent the two forces. The resultant force is marked. Since the two forces are at right angles the magnitude of the force can be calculated using Pythagoras. The angle of the resultant force can be found by measurement or calculation.

The resultant force is 61.8 N at 194°.

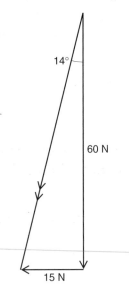

14°

60 N

15 N

Figure 6.17 ▲

Acceleration due to gravity

An object falling near the surface of the Earth accelerates (if the effects of air resistance are negligible). This acceleration is called the acceleration due to gravity (g).

Consider two bodies of mass 5 kg and 10 kg falling near the surface of the earth. We are assuming that there is no friction present.

Weight = force of gravity

$$= mg$$

$$= 5 \times 10$$

$$= 50 \text{ N}$$

weight = force of gravity

$$= mg$$

$$= 10 \times 10$$

$$= 100 \text{ N}$$

Since we are assuming no friction is present:

unbalanced force $= F_{un} =$ weight
$$= 50 \text{ N}$$

$$a = \frac{F_{un}}{m} = \frac{50}{5}$$

$$= 10 \text{ m/s}^2$$

unbalanced force $= F_{un} =$ weight
$$= 100 \text{ N}$$

$$a = \frac{F_{un}}{m} = \frac{100}{10}$$

$$= 10 \text{ m/s}^2$$

This shows clearly that the acceleration due to gravity in the absence of friction (air resistance) is the same for *all* bodies, no matter what their mass is.

The acceleration due to gravity on a planet has the same number value as the gravitational field strength on the planet. For example:

On a planet

acceleration due to gravity = gravitational field strength

On Jupiter   gravitational field strength = 25 N/kg

acceleration due to gravity = 25 m/s^2

Measuring the acceleration due to gravity

Using a light gate, computer and a piece of card cut as a mask we can calculate the acceleration due to gravity. The card is dropped through the light gate, from different heights, and the acceleration is measured in each case. It is found that the height the body is released from does not affect the acceleration.

For the Earth, the acceleration due to gravity $g = 10$ m/s^2.

Weight on Earth and other planets

The weight of an object near the surface of a planet such as the Earth is the pull of the planet on the object. Like all forces, the weight is measured in newtons.

The pull of gravity on a falling object (the weight) can be calculated using the equation

$$W = mg$$

In words, weight = mass × gravitational field strength

g on earth = 10 N/kg

The pull of gravity per kilogram mass on a planet is called its gravitational field strength. Gravitational field strength is measured in newtons per kilogram (N/kg).

Projectile motion

There are a wide variety of objects in orbit around the Earth, ranging in size from satellites the size of a house to an astronaut's glove. To understand how it is possible for these objects to stay in orbit, we study objects moving horizontally and vertically at the same time. These are called projectiles.

If an object is projected horizontally at 16 m/s then its inertia (unwillingness to change its motion) should tend to keep it moving at 16 m/s horizontally (in a straight line). Due to the force of gravity (the object's weight) this is not possible and the object experiences a force pulling it downwards while it tries to move horizontally at constant speed. To overcome this conflict between its inertia and its weight, the body follows a curved path (figure 6.18). This is called projectile motion. The motion of the projectile is therefore made up of two separate motions:

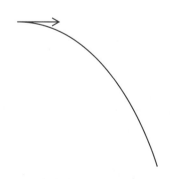

Figure 6.18 ▲
Projectile motion

1 A vertical motion under the influence of the force of gravity.

2 A horizontal motion under the influence of its inertia.

A special photograph is taken of a ball being projected, and a ball being dropped vertically at the same time, as shown in figure 6.19.

Figure 6.19 ▶
Stroboscopic photograph showing a ball being projected, and a ball being dropped vertically

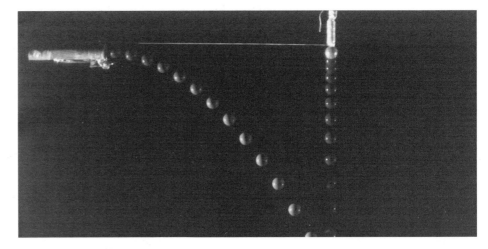

The vertical motion of the projectile and the free-falling object are in step all the way down. This means that the vertical motion of the projectile is the same as a free-falling object; that is, both are accelerating downwards with a constant acceleration of 10 m/s². This is the acceleration due to gravity, which was discussed earlier.

The horizontal spacings between the images of the projectile are all equal. This means that the horizontal motion of the projectile is constant speed and is equal to the speed of projection.

The motion of a projectile can be treated as two independent motions:

(a) Constant speed in the horizontal direction.
(b) Constant acceleration in the vertical direction due to the force of gravity.

The speed–time graphs for the two motions are shown in figure 6.20.

Figure 6.20 ▶
Motion of a projectile showing (a) constant speed in horizontal direction and (b) constant acceleration in vertical direction

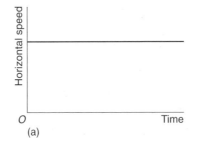 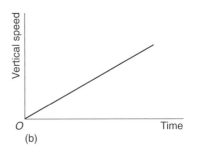

When solving problems, we can treat the motion in the horizontal direction *separately* from the motion in the vertical direction.

Example

A flare is fired horizontally out to sea from a cliff-top at a horizontal speed of 40 m/s. The flare takes 4 s to reach the sea.

(a) What is the horizontal speed of the flare after 4 s?
(b) Calculate the vertical speed of the flare after 4 s.
(c) Draw a graph to show how the vertical speed of the flare varies with time.
(d) Use this graph to calculate the height of the cliff top above the sea.

Solution

(a) Horizontal speed remains at 40 m/s.
(b) Using $v - u = at$ for the vertical motion, where $u = 0$,

$$a = g = 10 \text{m/s}^2$$

$$t = 4 \text{ s}$$

$$v - 0 = 4 \times 10$$

$$= 40 \text{ m/s}$$

(c) The graph is shown in figure 6.21.

(d) The height of the cliff is the area under the graph:

$$\text{distance} = \frac{1}{2} \times \text{base} \times \text{height}$$

$$= 0.5 \times 4 \times 40$$

$$= 80 \text{ m}$$

Height of the cliff above the sea is 80 m.

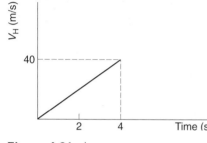

Figure 6.21 ▲

Newton's satellite

Newton, in considering the motion of the moon around the Earth, carried out the following 'thought experiment'. Suppose a bullet is fired horizontally from a gun situated on top of a high mountain. The bullet will have two motions, which occur simultaneously:

(a) A horizontal motion at uniform speed (if air resistance is negligible).
(b) A vertical motion of uniform acceleration downwards under the action of the gravitational attraction between the bullet and the Earth.

As a result, the bullet will follow a curved path and will hit the ground no matter how fast it was fired, as long as the Earth is flat.

However, the approximation of the 'flat' Earth is only valid over a limited range. For a 'round' Earth it becomes important to take the curvature of the Earth into account for projectiles of long range (figure 6.22).

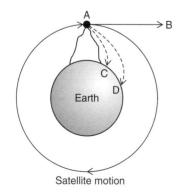

Figure 6.22 ▲

Motion of a bullet fired from a gun from the top of a high mountain

If there was no force of gravity, the bullet would follow the path AB, because if no forces act on it, it must travel with a uniform speed in a straight line (Newton's first law).

Due to the gravitational field of the earth, the bullet falls continuously below this line. If the bullet falls below the line *AB* faster than the Earth's surface curves away under it, the bullet will still hit the Earth, e.g. at the point *C* or *D*, depending on the bullet's speed.

If the bullet was fired at such a speed that it fell vertically at the same rate as the Earth's surface curved away under it, then it would always be at the same height above the surface. The bullet would then never reach the surface, but would circle it at a constant altitude. This was the first theory of the artificial satellite.

Remember, Newton 'thought' about this experiment long before we had rockets or satellites. In fact the height of Newton's fictional mountain would here to be about ten times that of Mount Everest.

Physics facts and key equations for forces

- Forces can change the shape, speed and direction of travel of an object.

- Mass is the amount of matter in a substance but weight is a force and is the Earth's pull on the object.

- Weight per unit mass is called the gravitational field strength.

- $W = mg$, where g is the gravitational field strength.

- The force of friction can oppose the motion of an object.

- Force is a vector.

- Equal forces acting in opposite directions on an object are called balanced forces and are equivalent to no force at all.

- Newton's first law states that an object will remain at rest or travel at constant velocity if the forces on the object are balanced.

- Newton's second law is $F = ma$.

- The resultant force is the one force which can replace all the forces acting on an object and have the same effect.

- Acceleration due to gravity = gravitational field strength.

- Projectile motion can be treated as two independent motions:

  horizontal motion – the object travels at constant velocity; vertical motion – the object has a constant downwards acceleration.

End-of-Chapter Questions

1. A motorcyclist reaches a speed of 20 m/s and this is the maximum speed that can be reached.

   Draw a diagram showing and naming the forces acting on the bike and describe why the forces are balanced.

2. Use the data sheet to calculate the weight of a person of mass 60 kg standing on the surface of the following: **(a)** Earth **(b)** Moon **(c)** Mars.

3. Copy and complete the table shown below

Force in N	Mass in kg	Acceleration in m/s²
	70	5
350	105	
5000		2.5

4. A car has a mass of 800 kg and accelerates at 3.5 m/s². Calculate the unbalanced force on the car.

5. A car is being pushed on a cold day to start it moving. The car is moving at a constant speed. Name the force opposing the pushing force.

6. The force acting on an electric toy car due to the engine is 200 N. The frictional force acting against it is 30 N.

   (a) What is the value of the unbalanced force acting on the car?

   (b) Calculate the acceleration of the car if it has a mass of 25 kg.

7. A luggage trolley is being pulled by two ropes. Both ropes have forces of 75 N acting on them. If the two ropes are replaced by one rope which has to have exactly the same effect, what is this replacement force called?

8. The acceleration due to gravity on a planet is 26 m/s². What can you state about the gravitational field strength on this planet?

9. An aircraft is moving at 55 m/s when it drops supplies to help in a relief operation. The package takes 8 s to reach the ground.

   (a) Calculate how far the aircraft has travelled in this time.

   (b) What is the speed of the package just before reaching the ground if the parachute does not open?

7 Momentum and energy

After reading this chapter you should be able to:

1 State Newton's third law.

2 Identify newton pairs in situations involving several forces.

3 State that momentum is the product of mass and velocity.

4 State that momentum is a vector quantity.

5 State that the law of conservation of linear momentum can be applied to the interaction of two objects moving in one direction, in the absence of external forces.

6 Carry out calculations in which all the objects move in the same direction and one object is initially at rest.

7 State that work done is a measure of the energy transferred.

8 Carry out calculations involving the relationship between work done, force and distance.

9 State the relationship and carry out calculations involving the relationship between power, work and time.

10 Carry out calculations between gravitational potential energy and mass, gravitational field strength and change in height.

11 Carry out calculations involving the relationship between kinetic energy, mass and speed.

12 Carry out efficiency calculations involving output power/output energy and input power/input energy.

Newton's third law

When you try and strike a ball using a bat, the ball moves because a force is exerted on it. But you will also feel a force through the bat. If you miss the ball then there is no force on anything. Forces only exist in pairs which are equal in size but opposite in direction.

This is Newton's third law. It is sometimes written as:

Action and reaction are equal and opposite

In figure 7.1 a parcel is placed on a table. The parcel weighs 500 N and this is the force that the parcel exerts downwards on the table top. The table top exerts a force of 500 N upwards on the parcel.

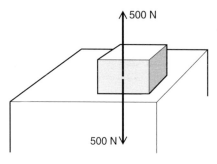

Figure 7.1 ▲
Forces acting on a parcel resting on a table

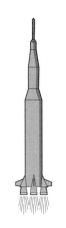

Figure 7.2 ▲
Forces acting on a rocket

Jet engines in aircraft and rocket propulsion use this principle. In both cases a high-speed stream of hot gases (produced by burning fuel) is pushed backwards from the vehicle with a large force (figure 7.2) and a force of the same size pushes the rocket forwards.

Other examples:

- If you push against a wall, the wall pushes you backwards.
- In walking, you push your foot backwards but the floor pushes you forwards.
- In swimming, you push the water backwards but you are pushed forward.

Put simply, this means that when *A* pushes *B*, *B* pushes *A* back with the same size of force.

Rockets and jet engines are able to produce motion.

Action: *vehicle* pushes *hot gases backwards* (downwards)

Reaction: *hot gases* push *vehicle forwards* (upwards)

Collisions and momentum

When a car collides with an object such as a stationary car the speed before the collision can sometimes be calculated from the tyre marks on the road. If two cars of different mass hit an identical stationary object then we would expect the heavier car to have the greater damage. But what if the lighter car was moving faster?

To analyse these problems we use a concept called momentum:

momentum = mass × velocity

        units = kg m/s

Momentum is a vector quantity.

The basic law of all collisions is that

total momentum before collision = total momentum after the collision

provided no external forces are acting.

This concept allows us to analyse simple collisions without knowing the forces that act during the collision.

Example
A trolley of mass 10 kg is moving to the right at 6 m/s. It collides with another stationary trolley of mass 20 kg. After the collision both trolleys move off together. Calculate the velocity with which they move (figure 7.3).

Figure 7.3 ▶

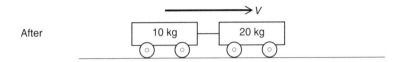

Solution

Before the collision, total momentum $= 10 \times 6 = 60$ kg m/s

After the collision, total momentum $= (10 + 20)\, v = 30\, v$

where v is the common velocity after the collision.

Since total momentum before $=$ total momentum after,

$$60 = 30\, v$$

$$v = 2 \text{ m/s}$$

Work and energy

Energy is a very useful quantity. Simply, energy allows objects to move and lets you do things: it is called work.

Although energy has many different forms, there are ways of measuring them. When work is done, energy is transferred to an object or changed into another form.

energy transferred $=$ work done
$=$ force $\times$ distance moved by the force

Thus $\text{Wd} = F \times d$

where d distance in metres and $F =$ force in newtons.

The unit of energy and work done is the joule (J); thus

$$1 \text{ joule} = 1 \text{ newton metre}$$

$$1 \text{ J} = 1 \text{ N m}$$

Example

A boy finds he has to exert a force of 50 N to lift a box 2 m onto a shelf.

Solution

$$Wd = F \times d$$

where $F = 50$ N and $d = 2$ m

$$Wd = 50 \times 2$$
$$= 100 \text{ J}$$

Conservation of energy

Energy cannot be created or destroyed, but it can be changed from one form to another, when work is done. For example, the kinetic energy of movement in a car is changed into mainly heat when the brakes are applied. If a system gains energy then another system loses the same amount.

This is called the conservation of energy.

Gravitational potential energy (E_p)

Water in a mountain loch has stored energy (gravitational potential energy) which can be transferred into electrical energy in a hydroelectric scheme. It is available as the water is above the generating station and can be transferred by allowing it to fall. We can derive an equation to allow us to calculate the potential energy for a mass being lifted.

A mass m is lifted at constant speed through a vertical height of h metres (figure 7.4). The work done in lifting it is calculated as

$$\text{work done} = F \times d$$

In this case the force applied must balance the weight of the box.

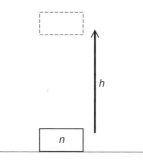

Figure 7.4 ▲
A mass being lifted vertically at constant speed

applied force upwards = weight downwards
                      $= mg$
Work done on object = $mg \times h$
                    $= mgh$
                    = gain in gravitational potential energy (E_p)
      $E_p = mgh$

where

E_p = change in gravitational E_p (J),
m = mass (kg),
g = gravitational field strength (N/kg) and
h = vertical height (m).

Kinetic Energy (E_k)

This is the energy possessed by any object which moves. We can also derive an equation for this type of energy.

A mass m starts from rest and is accelerated by a force F to a final speed v in time t (figure 7.5).

Figure 7.5 ▶
A mass being accelerated from rest

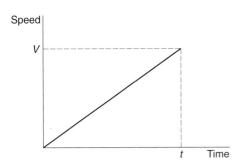

Figure 7.6 ▲

Work done on object $= F \times d$

$= ma \times$ area under speed–time graph (figure 7.6)

and $a = \dfrac{v-u}{t} = \dfrac{v-0}{t} = \dfrac{v}{t}$

area under the graph $= \dfrac{1}{2} \times t \times v = \dfrac{1}{2} vt$

work done on object $= \dfrac{mv}{t} \times \dfrac{1}{2} vt$

$= \dfrac{1}{2} mv^2$

work done on object = gain in kinetic energy (E_k)

$$E_k = \dfrac{1}{2} mv^2$$

Note that E_k depends on v^2. This means that:

- If the speed doubles then the kinetic energy increases by four.
- If the speed trebles then the kinetic energy increases by nine.

Power

When we say that one appliance is more powerful than another device then we mean that it uses up energy at a faster rate.

power = energy transferred in 1 s = work done in 1 s

$$\text{power} = \frac{\text{energy transferred}}{\text{time taken}} = \frac{\text{work done}}{\text{time}}$$

Power is measured in watts (W). A power of one watt means that one joule of energy is transferred in one second.

Example 1

A 3 kg box is raised through 20 m in 4 s.

(a) Find the gain in gravitational E_p of the box.
(b) Find the power used to lift the box.
(c) Why would more power be needed than the value calculated in (b)?

Solution

(a) $E_p = mgh$

$$= 3 \times 10 \times 20$$

$$= 600 \text{ J}$$

(b) $P = \dfrac{E}{t}$

$$= \dfrac{600}{4}$$

$$= 150 \text{ W}$$

(c) Some of the energy supplied is changed into heat and is not used to lift the box.

Example 2

A 50 kg girl on a 15 kg bicycle is moving at a uniform speed of 5 m/s. She applies the brakes and comes to rest in 2 seconds.

(a) What is the kinetic energy of the girl and her bicycle before she brakes?
(b) What becomes of this kinetic energy during braking?
(c) Calculate the power of the brakes.

Solution

(a) $E_k = \dfrac{1}{2} mv^2$

$$= \dfrac{1}{2} \times 65 \times (5)^2$$

$$= 813 \text{ J}$$

(b) It changes into heat.

(c) Change in $E_k = 813 \text{ J} =$ energy transferred.

$$P = \dfrac{E}{t}$$

$$= \dfrac{813}{2}$$

$$= 407 \text{ W}$$

Figure 7.7 ▲
Estimating your own power

Estimating your own power

In going upstairs, you will gain in gravitational potential energy. An applied force equal to your weight (force of gravity) acts vertically upwards and so the distance moved by the applied force is the vertical height of the stairs.

Measure your mass on bathroom scales (figure 7.7). To calculate your power you also need the time taken for you to go up the stairs. Run up the stairs and time how long it takes you to run up a measured vertical height.

Typical results

height of stairs $= 3$ m

time taken to go up stairs $= 3.5$ s

weight $= 700$ N

$$
\begin{aligned}
\text{work done going up stairs} \ &= \text{energy transferred} \\
&= \text{gain in gravitational potential energy} \\
&= mgh \\
&= 700 \times 3 \\
&= 2100 \text{ J}
\end{aligned}
$$

$$
\text{power} = \frac{\text{energy transferred}}{\text{time taken}} = \frac{mgh}{t} = \frac{2100}{3.5}
$$
$$
= 600 \text{ W}
$$

The power of a horse, on the other hand, is about 750 W. An Olympic runner can produce up to 3000 W but we cannot sustain this power for any length of time. A typical athlete will lose up to 50% of the energy produced within the body as heat.

Energy conservation: gravitational E_p to E_k

During any energy transformation the total amount of energy is always conserved; that is, it stays the same, but may be changed into less useful forms.

Energy transfer or energy changes are associated with a falling object (Figure 7.8).

As the object falls it 'loses' gravitational E_p but gains E_k. Just before impact with the ground it has 'lost' all its gravitational E_p and has only E_k (figure 7.8).

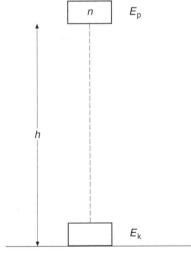

Figure 7.8 ▲
A falling body

If no energy is transferred with the surroundings, then

gain in E_k = loss in gravitational E_p

$$\frac{1}{2}mv^2 - 0 = mgh$$

$$\frac{1}{2}mv^2 = mgh$$

Example

The Petronas Tower in Kuala Lumpur, Malaysia is 750 m in height. John is visiting on holiday and drops his bag over the edge. What is its speed on reaching the ground? (See figure 7.9)

Figure 7.9 ▶
The Petronas Tower, Kuala Lumpur

Solution

The change in gravitational potential energy equals the change in kinetic energy. Then

$$mgh = \frac{1}{2}mv^2 - 0$$

The masses will cancel on both sides of the equation, giving

$$gh = \frac{1}{2}v^2$$

The height and speed are connected but are independent of mass. All masses will fall and hit the ground at the same time.

$$10 \times 750 = \frac{1}{2}v^2$$

$$v^2 = 20 \times 750$$

$$= 15\,000$$

$$v = 122 \text{ m/s}$$

Efficiency

In any machine there is always some energy lost to another form. This occurs, for example, as work done against friction, which will produce heat in a motor car engine or any electric motor. While the total energy of the system cannot change, the useful energy output can be compared to the useful energy input. This is called efficiency.

$$\text{efficiency} = \frac{\text{useful energy output}}{\text{total energy input}} \times 100\%$$

In a similar way we can express the efficiency in terms of power:

$$\text{efficiency} = \frac{\text{useful power output}}{\text{total power input}} \times 100\%$$

Most machines have a range of efficiencies.

For a motor car efficiency is about 20%. That is, only 20% of the energy input from the petrol is converted into kinetic energy; the remainder goes as heat either into the engine block or into the atmosphere as exhaust gases. While this may seem very low and inefficient we tolerate it because only petrol or diesel can give us the rapid acceleration which we demand in moving vehicles.

Example

An electric motor has an input power of 2.0 kW. The output power is 1.4 kW. Calculate the efficiency of this motor.

Solution

$$\text{Efficiency} = \frac{\text{useful power output}}{\text{total power input}} \times 100\%$$

$$= \frac{1.4}{2.0} \times 100\%$$

$$= 70\%$$

No machine can reach or exceed 100%, because this would mean it was creating energy, which is impossible, and so provides a check on your calculations.

- Newton's third law states that forces always act in pairs which are equal in size but opposite in direction.

- Momentum = mass × velocity.

- Momentum is a vector quantity and the units are kg m/s.

- Total momentum before a collision = total momentum after, in the absence of external forces.

- Work done = force × distance and is measured in joules.

- Power $P = \dfrac{\text{work done}}{\text{time}}$ and is measured in watts.

- Potential energy $E_p = mgh$

- Kinetic energy $E_k = \dfrac{1}{2}\,mv^2$

- Efficiency $= \dfrac{\text{useful work output}}{\text{total work input}} \times 100\%$

End-of-Chapter Questions

1 A person stands in a lift which moves upwards at constant speed. What can you state about the forces acting on the person and the floor of the lift?

2 In Figure 7.10 block B has a mass of 20 kg and is at rest. Block A has a mass of 10 kg and is moving at a speed of 6 m/s.

(a) Calculate the momentum of block A before the collision.

(b) Calculate the combined speed of the two objects if they stick together after the collision.

Figure 7.10 ▲

3 A wagon is moved by a force of 20 N. It travels a distance of 30 m.

Calculate the work done on the wagon.

4 Copy and complete the table below.

Force in N	Distance moved in m	Work done in J
50	24	
	125	2500
90		1800

5 A force of 10 N is exerted on a supermarket trolley and it is moved a distance of 30m. If this happens in 5 s, calculate the power.

6 Copy and complete the table below.

Mass in kg	Acceleration due to gravity in m/s²	Height raised in m	Potential energy in J
5	10	8	
25	10		1250
90	1.6		1440
50		2	2600

7 Oranges hang from a branch on a tree. An orange has a mass of 200 g and is at a height of 7 m.

Calculate the potential energy of the orange.

continued ➤

8 A 20 000 kg rocket is travelling upwards at 1000 m/s. Calculate the kinetic energy of the rocket.

9 Copy and complete the table below.

Mass in kg	Speed in m/s	Kinetic Energy in J
5	10	
20		2250
	13	1014

10 A ball of mass 0.25 kg falls from a height of 3.2 m.

(a) Calculate its potential energy.

(b) Assumimg that all the potential energy is transferred to kinetic energy just before it reaches the ground, calculate its speed on reaching the ground.

8 Heat

At the end of this chapter you should be able to:

1 State that the same mass of different materials need different quantities of heat energy to change their temperature by one degree Celsius.

2 Carry out calculations involving specific heat capacity.

3 State that heat is gained or lost by a substance when its state is changed.

4 State that a change of state does not involve a change of temperature.

5 Carry out calculations involving specific latent heat.

6 Carry out calculations involving energy, work, power and the principle of conservation of energy.

Heat and temperature

Heat, just like light and sound, is a form of energy and is measured in joules (J). Temperature is a measure of how hot a substance is and is measured in degrees Celsius (°C).

Heat transfer

Heat is always transferred from a higher temperature to a lower temperature. There are three possible ways in which it can be transferred called conduction, convection and radiation.

Conduction
The heat is transferred through a solid material. The particles making up the solid cannot change their position but pass the heat from particle to particle. Heat moves from the high temperature to the low temperature. Materials which allow heat to move easily through them are called conductors – metals are the best conductors. Materials which do not allow heat to move through them easily are called insulators – liquids and gases are good insulators (poor conductors).

Convection
The heat is transferred by the movement of the heated particles making up the liquid or gas. The heated fluid (liquid or gas) becomes less dense and rises, and cold fluid falls to take its place, that is, convection currents are set up. Convection cannot take place in a solid because the particles cannot move away.

Many heat insulators contain trapped air, e.g. cotton wool, felt, woollen clothes. No convection can take place because the air is trapped and cannot move. Air is also a non-metal, so it is a poor conductor of heat.

Radiation

Radiation travels in straight lines until absorbed by an object. It can travel through a vacuum (the Earth is heated by radiation from the sun). Heat radiation travels at a speed of 3×10^8 m/s. All hot materials radiate heat energy.

Changing temperature

When you want to make a hot drink you put some water in the kettle and switch it on. The water will get hot as a result of the heat energy supplied by the heating element of the kettle. However, the amount of heat energy required to warm the water depends on:

- The temperature rise – more energy is needed for a larger temperature rise.
- The mass of the water – more energy is needed for a greater mass of water.

Example

It requires 20 900 J of energy to increase the temperature of 0.5 kg of water by 10°C.

(a) How much energy is required to increase the temperature of 0.5 kg of water by 20°C?
(b) How much energy is required to increase the temperature of 1 kg of water by 40°C?

Solution

(a) It takes 20 900 J to change the temperature of 0.5 kg of water by 10°C, so it will take twice as much energy to change the temperature of 0.5 kg by 20°C. Therefore it takes 41 800 J to change the temperature of 0.5 kg of water by 20°C.
(b) It takes 20 900 J to change the temperature of 0.5 kg of water by 10°C.
It takes 41 800 J to change the temperature of 1 kg of water by 10°C.
It takes 167 200 J to change the temperature of 1 kg of water by 40°C.

Specific heat capacity

Equal masses of water and copper are supplied with the same quantity of heat energy. It is found that the copper has a higher rise in temperature.

The heat energy needed to change the temperature of a substance depends on:

- The change in temperature of the material (ΔT).
- The mass of the material (m).
- The type of material (specific heat capacity, c, of the material).

The specific heat capacity of a substance is the amount of energy required to change the temperature of 1 kg of a substance by 1°C. The units of specific heat capacity are joules per kilogram per degree Celsius (J/kg °C).

Water has a specific heat capacity of 4180 joules per kilogram per degree Celsius (4180 J/kg °C). This means that it takes 4180 joules of heat energy to change the temperature of 1 kg of water by 1°C; that is, it would take 4180 J to change the temperature of 1 kg of water by 1°C, so it would take 16 720 J to change the temperature of 4 kg of water by 1°C, and it would take 83 600 J to change the temperature of 4 kg of water by 5°C.

The energy needed to change the temperature of a substance can be put in the form of an equation:

$$E_h = cm\Delta T$$

where E_h = energy needed to change the temperature (J), c = specific heat capacity (J/kg °C), m = mass of substance (kg) and ΔT = change in temperature (°C).

Example

How much energy is required to change the temperature of 0.5 kg of water from 18°C to 100°C? The specific heat capacity of water is 4180 J/kg °C.

Solution

$E_h = cm\Delta T$

$E_h = 4180 \times 0.5 \times (100 - 18)$

$E_h = 4180 \times 0.5 \times 82$

$E_h = 1.71 \times 10^5$ J

Example

A 750 g steel cooking pan cools from 65°C to 18°C. How much heat energy is released by the pan? The specific heat capacity of steel is 500 J/kg °C.

Solution

$$E_h = cm\Delta T$$

$$E_h = 500 \times 0.750 \times (65 - 18)$$

$$E_h = 500 \times 0.750 \times 47$$

$$E_h = 1.76 \times 10^4 \text{ J}$$

Heat problems

Energy can be changed from one form to another. The total amount of energy remains unchanged – this is the principle of conservation of energy. An electric heater converts electrical energy into an equal amount of heat energy.

Due to conduction, convection and radiation, some of the heat energy supplied by the heater will be transferred ('lost') to the surroundings. This means that the substance will absorb (take in) less energy than was supplied by the heater.

energy supplied = energy absorbed + energy transferred to the surroundings

In most heat problems it is assumed that no energy is transferred to the surroundings. Hence:

energy supplied = energy absorbed by the material

Example

A well-insulated kettle is switched on for 2 minutes. The kettle heats 0.8 kg of water. The temperature of the water changes from 16°C to 100°C. Calculate the power rating of the kettle. The specific heat capacity of water is 4180 J/kg °C.

Solution

$$E_h = cm\Delta T = 4180 \times 0.8 \times (100 - 16)$$

$$E_h = 4180 \times 0.8 \times 84 = 280\ 896 \text{ J}$$

$$P = \frac{E}{t} = \frac{280\ 89}{2 \times 60} = 2341 \text{ W} = 2.3 \text{ kW}$$

Example

A deep fat fryer is used to heat 800 g of cooking oil of specific heat capacity 3000 J/kg °C. The temperature of the cooking oil changes from 20°C to 140°C in a time of 180 s.

(a) Calculate the power rating of the deep fat fryer.

(b) The deep fat fryer operates from the 230 V mains. Find the current in the element of the deep fat fryer.

Solution

(a) $E_h = cm\Delta T = 3000 \times 0.8 \times (140 - 20)$

$E_h = 3000 \times 0.8 \times 20 = 288\,000\text{ J}$

$$P = \frac{E}{t} = \frac{288\,000}{180} = 1600\text{ W} = 1.6\text{ kW}$$

(b) $P = IV$

$1600 = I \times 230$

$$I = \frac{1600}{230} = 6.96\text{ A}$$

Specific latent heat

When cold water in a kettle is heated its temperature rises until the water starts to boil at 100°C. Further heating of the water no longer produces a rise in temperature of the water but steam at 100°C is produced. The heat energy supplied by the kettle is now being used to change water at 100°C into steam at 100°C. The energy required to change 1 kg of a liquid at its boiling point into 1 kg of vapour at the same temperature is called the specific latent heat of vaporisation. The word **latent** means hidden and refers to the fact that the temperature of the substance does not change and the heat energy supplied to the substance seems to have disappeared.

When ice at its melting point of 0°C is heated it turns into water at 0°C. Heat energy is required to change the ice to water without a change in temperature. The energy required to change 1 kg of a solid at its melting point into 1 kg of liquid at the same temperature is called the specific latent heat of fusion.

The three states of matter are solid, liquid and gas. Whenever a substance changes state, latent heat energy is required.

When a substance changes from a solid to a liquid or a liquid to a gas, energy is needed to break down the force (or bond) holding the particles together and to push the particles further apart.

- Specific latent heat of fusion: the energy required to change 1 kg from solid at its melting point to liquid without change in temperature.
- Specific latent heat of vaporisation: the energy required to change 1 kg from liquid at its boiling point to gas without change in temperature.

When a substance changes state from solid to liquid or liquid to gas, latent heat is absorbed (taken in). When a substance changes state from gas to liquid or liquid to solid, latent heat is released (given out). When a substance changes state there is no change in temperature.

The symbol for specific latent heat is l, and it is measured in J/kg.

The specific latent heat of a material is the heat energy required to change the state of 1 kg of the material without a change in temperature.

For a material with a specific latent heat l:

- To change the state of 1 kg of the material at constant temperature requires l J.
- To change the state of m kg of the material at constant temperature requires $m \times l$ J.

$$\text{That is, } E_h = ml$$

where E_h = energy needed to change state (J); m = mass which changed state (kg); l = specific latent heat (J/kg)

Example

A heater with a power rating of 50 W is placed in a well-insulated bath of 0.6 kg of ice at 0°C. How long does the heater require to be switched on to melt 0.05 kg of the ice? (Assume that no ice is melted due to room temperature and that all the heat energy from the heater is absorbed by the ice.) The specific latent heat of fusion of water is 3.34×10^5 J/kg.

Solution

$$\text{energy required to melt ice } = ml$$
$$= 0.05 \times 3.34 \times 10^5$$
$$= 1.67 \times 10^4 \text{ J}$$

$$\text{energy supplied by heater } = \text{energy required to melt ice}$$
$$\text{P} \times \text{t} = 1.67 \times 10^4$$
$$50 \times \text{t} = 1.67 \times 10^4$$
$$t = 334 \text{ s}$$

Example

During an experiment a teacher does not put the lid on a kettle, so that the automatic cutout does not work when the water reaches a temperature of 100°C. The teacher recorded the following information while the water was boiling: mass of water changed to steam = 100 g, power rating of kettle = 2200 W, time of supply = 105 s.

(a) Calculate the specific latent heat of vaporisation of water obtained from this experiment.

(b) Explain why this value is likely to be higher than the accepted value for the specific latent heat of vaporisation of water.

Solution

(a) Energy supplied by heater $E_h = P \times t = 2200 \times 105 = 231\ 000$ J.

$$E_h = ml$$

$$231\ 000 = 0.1\ l$$

$$l = 2\ 310\ 000\ \text{J / kg}$$

(b) Not all of the energy supplied by the heater element is absorbed by the water. Some of it is 'lost' in heating up the air around the kettle. This means that more energy is supplied to produce 0.1 kg of steam and so the value for *l* is too large.

 Physics facts and key equations for heat

- Equal masses of different substances require different amounts of energy to change their temperature by 1°C.

- During a change in temperature the energy absorbed or lost by a substance, E_h, is measured in joules (J), specific heat capacity, c, is measured in joules per kilogram per degree Celsius (J/kg °C), mass, m, is measured in kilograms (kg) and the change in temperature, ΔT, is measured in degrees Celsius (°C).

- Energy absorbed or lost = specific heat capacity × mass × change in temperature:

$$E_h = cm\Delta T$$

This equation is used whenever there is a change in temperature of the substance.

- A specific heat capacity, c, of 100 J/kg °C means that 100 J of energy is required to change the temperature of 1 kg of the substance by 1°C.

- A change of state occurs when a solid changes into a liquid (or a liquid changes to a solid) or a liquid changes to a gas (or a gas changes to a liquid).

- There is no change in temperature when a change of state occurs.

- During a change in state, the energy absorbed or lost by the substance, E_h, is measured in joules (J), mass, m, is measured in kilograms (kg) and the specific latent heat of fusion or vaporisation, l, is measured in joules per kilogram (J/kg).

- Energy absorbed or lost = mass × specific latent heat

$$E_h = ml$$

This equation is used whenever there is a change in state of the substance.

End-of-Chapter Questions

1 It takes 8360 J of heat energy to raise the temperature of 2 kg of water by 1°C. How much heat energy will be required to raise the temperature of: (a) 4 kg of the water by 1°C; (b) 4 kg of the water by 5°C; (c) 8 kg of the water by 10°C?

2 Calculate the amount of heat energy required to raise the temperature of

   (a) 0.5 kg of water from 18°C to 58°C? (the specific heat capacity of water is 4180 J/kg °C).
   (b) a 0.95 kg steel baking tray from 18°C to 198°C (the specific heat capacity of steel is 500 J/kg °C).

3 A well-insulated kettle contains 1.2 kg of water at a temperature of 20°C. The kettle is switched on. After 180 s the water reaches a temperature of 100°C. The specific heat capacity of water is 4180 J/kg °C

   (a) How much energy is absorbed by the water in 180 s?
   (b) Calculate the power rating of the kettle.

4 A heater operating from a 12 V supply draws a current of 4 A. The heater is used to heat a 1 kg copper block. The heater is switched on for 5 minutes (the specific heat capacity of copper is 386 J/kg °C).

   (a) How much heat energy was produced by the heater in 5 minutes?
   (b) Calculate the maximum possible rise in the temperature of the copper block.
   (c) Explain why the rise in temperature of the copper block will be less than the answer to (b).

5 How much energy is required to change 0.8 kg of ice at 0°C into water at 0°C? (the specific latent heat of fusion of ice is 3.34×10^5 J/kg).

6 How much energy is required to change 0.2 kg of steam at 100°C into water at 100°C? (the specific latent heat of vaporisation of water is 2.26×10^6 J/kg).

7 A 2000 W heater is used to bring 1.5 kg of water to its boiling point. Calculate the mass of water boiled off if the heater is left on for a further 80 s (the specific latent heat of vaporisation of water is 2.26×10^6 J/kg).

1 A train travels due north for 25 km and then west for 60 km.

   (a) Calculate the speed in m/s if the journey takes 40 minutes.

   (b) Explain why the velocity has a lower value than the speed and what additional quantity must be stated to give the velocity.

   (c) During part of this journey the train increases its speed from 14 m/s to 28 m/s in 3.5 s.

Calculate the acceleration of the train.

2 During a game a ball of mass 0.15 kg is struck by a bat and accelerates at 6.5 m/s^2.

   (a) Calculate the force acting on the ball.

   (b) The ball reaches a speed of 70 m/s. Calculate the kinetic energy of the ball.

   (c) The ball strikes the ground. State what happens to the kinetic energy upon impact.

3 A car of mass 1000 kg is moving at 8 m/s when it strikes another stationary car of mass 600 kg.

   (a) If both cars lock together on impact, calculate the common speed after the collision.

   (b) In designing these cars, the drag forces on the car are reduced. Describe two features of car design which will produce these effects.

4 An ice tray contains 0.15 kg of water. The initial temperature of the water is 20°C. The ice tray and the water are placed in a freezer. The freezer is at a constant temperature of –19°C. Show that the freezer has to remove 68.6 kJ of heat energy in order to change 0.15 kg of water at 20°C into ice at –19°C (the specific heat capacity of water = 4180 J/kg °C, the specific latent heat of fusion of water = 3.34×10^5 J/kg, the specific heat capacity of ice = 2100 J/kg °C).

Waves and Optics

9 Waves

At the end of this chapter you should be able to:

1 State that a wave transfers energy.

2 Describe a method of measuring the speed of sound in air, using the relationship between distance, time and speed.

3 State that radio and television signals are transmitted through air at 300 000 000 m/s and that light is also transmitted at this speed.

4 Carry out calculations involving the relationship between distance, time and speed in problems on water waves, sound waves, radio waves and light waves.

5 Use the following terms correctly in context: wave, frequency, wavelength, speed, amplitude and period.

6 State the difference between a transverse and a longitudinal wave and give an example of each.

7 Carry out calculations involving the relationship between speed, wavelength and frequency for waves.

8 State, in order of wavelength, the members of the electromagnetic spectrum: gamma rays, X-rays, ultraviolet, visible light, infrared, microwaves, TV and radio.

Waves

Waves such as water waves are easily visible. We can see them breaking on the shore and the effect of large waves can cause considerable damage during storms or typhoons (figure 9.1).

Figure 9.1 ▶
Waves can carry a great deal of energy!

Waves transfer energy from one point to another. In the case of water waves this transfer can be seen since wood and other materials are often washed onto beaches. However, sound, light and radio and TV waves are also methods of transferring energy.

We see a lightning flash before we hear the thunder. This tells us that light travels faster than sound, but how fast does sound travel? To answer this question we need to measure the speed of sound.

Measuring the speed of sound

This requires the equation met earlier that speed $= \dfrac{\text{distance}}{\text{time}}$

$$v = \frac{d}{t}$$

In a material the speed of sound is constant.

An accurate method uses a computer as a timer to measure the time for a sound signal to travel from one microphone to another (figure 9.2). You could do this without an electronic timer but reaction time could be important if the time interval is very small.

Figure 9.2 ▶
Measuring the speed of sound

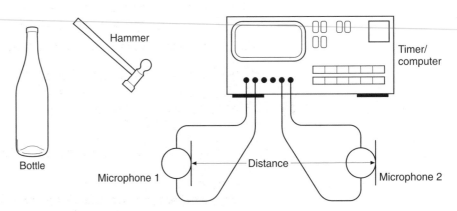

The microphones are placed at least 1 m apart. The distance will be measured precisely with a metre rule. When a loud sound is made by hitting a bottle, the first microphone starts the timing. When the sound reaches the second microphone the timer stops timing.

Typical results
distance between the microphones $= 2.0$ m

time on the timer $= 6$ ms $= \dfrac{6}{1000} = 0.006$ s (ms is a millisecond, which is 1/1000 s)

speed of sound $= \dfrac{\text{distance travelled}}{\text{time taken}} = \dfrac{2.0}{0.006}$

value of speed of sound $= 333$ m/s

In air the speed of sound will vary due to temperature and the value given in the data table is 340 m/s.

The speed of light

Radio and television signals travel about one million times faster than sound. These waves and light waves travel in air at 300 million m/s. This value for the speed of light is now the constant against which all other physical measurements can be found.

Properties of waves

All waves have certain properties, which are shown in figure 9.3. The top part of the wave is called the **crest** and the bottom part is the **trough**. The line running through the middle of the wave pattern is called the **axis**.

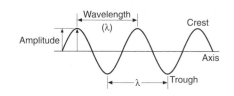

Figure 9.3 ▲
Properties of a wave

The distance from the axis to the top of the crest or the bottom of the trough is called the **amplitude**. The **wavelength** is the distance after which the pattern repeats itself. It is given the symbol λ (lambda).

All distances are measured in metres or occasionally centimetres.

The **frequency** of the wave is the number of waves per second and is measured in hertz (Hz). The **period** is the time for one complete wave to travel past a point and is measured in seconds.

The link between frequency f and time period T is given as

$$T = \frac{1}{f}$$

If the frequency is known the period can be found and vice versa.

Example
A light signal is sent a distance of 5 km in air. Calculate the time for this journey.

Solution

Using the equation $v = \dfrac{d}{t}$

$v = 300\ 000\ 000$ m/s and $5 = 5000$ m

$$t = \frac{5000}{300\ 000\ 000}$$

$$= 1.7 \times 10^{-5} \text{ s}$$

This very small time shows that if TV and radio are broadcast at this speed then we can receive instantaneous pictures of events from around the world.

Transverse and longitudinal waves

If we take a slinky spring then it is possible to create two different types of waves.

If you push the end of the spring back and forth then the coils become alternately compressed and then extended. You can also see the wave running along the coil. This type of wave is called a longitudinal wave (figure 9.4). In this type of wave the particles vibrate back and forwards along the direction the wave is travelling.

Figure 9.4 ▶
A longitudinal wave

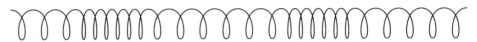

If you move the end of the spring from side to side then each coil moves from side to side as the wave travels along the spring (figure 9.5). This is called a transverse wave, in which the particles vibrate at right angles to the direction of motion of the wave.

Figure 9.5 ▶
A transverse wave

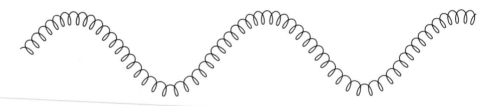

Sound waves are longitudinal waves. Light and all forms of electromagnetic waves are transverse waves. Some waves have different components. Seismic waves produced from the Earth's core and water waves are complex forms of both transverse and longitudinal.

The speed of the wave can be found from the usual equation:

$$\text{average speed} = \frac{\text{distance travelled}}{\text{time taken}}$$

$$v = \frac{d}{t}$$

where v = speed in metres per second, d = distance travelled in metres and t = time taken in seconds.

The speed of the wave can also be found from the following equation. In words:

$$\text{speed} = \text{frequency} \times \text{wavelength}$$

$$v = f \times \lambda$$

In units: m/s = hertz × metres

This means that there are two equations for finding the speed of waves.

Wave calculations

We can use the wave equations to find some of the features of waves.

Example

Waves of frequency 6 Hz are sent down a rope. The speed of the waves is 12 m/s.

Calculate the wavelength and the period of the wave.

Solution

Using $v = f \times \lambda$

$$f = 6 \text{ Hz and } v = 12 \text{ m/s}$$

$$\lambda = \frac{v}{f} = \frac{12}{6} = 2 \text{ m}$$

$$T = \frac{1}{f} = \frac{1}{6} = 0.17 \text{ s}$$

Example

A radio station broadcasts on a frequency of 198 kHz. What is the wavelength of this station?

$$v = 3 \times 10^8 \text{ m/s}$$

and $\quad f = 198 \times 10^3 \text{ Hz}$

$$\lambda = \frac{v}{f} = \frac{3 \times 10^8}{198 \times 10^3} = 1520 \text{ m}$$

Electromagnetic spectrum

The waves listed below travel at 3×10^8 m/s (300 000 000 m/s). Each radiation has a different wavelength and frequency and will need a specific detector. Each radiation is shown with a suitable detector.

Radiation	Detector
Smallest wavelength and highest frequency Gamma rays (from radioactive substances such as cobalt 60 or certain rocks or power stations)	Photographic film or Geiger counter
X-rays (from X-ray tubes)	Photographic film
Ultraviolet (sun or certain lamps)	Skin or film which is said to fluoresce
Visible light (sun)	Eye or photographic film
Infrared (lamps or very hot objects)	Photo transistor
Microwaves (cookers)	Aerial
TV and radio (transmitters) *Largest wavelength and lowest frequency*	Aerial

As the wavelength increases the frequency decreases but the product of frequency and wavelength is a constant, namely the speed of light.

Physics facts and key equations for waves

- A wave transfers energy.

- Radio, television signals and light are transmitted through air at 300 000 000 m/s.

- To calculate the various features of waves use

$$v = f \times \lambda \text{ or } v = \frac{d}{t}$$

- In a transverse wave energy travels at right angles to the direction of the wave, an example being light.

- In a longitudinal wave the wave energy travels parallel to the wave direction, an example being a sound wave.

- The electromagnetic spectrum, from short to long wavelengths, is in this order: gamma rays, X-rays, ultraviolet, visible light, infrared, microwaves, TV and radio waves.

End-of-Chapter Questions

1. In an experiment to measure the speed of sound, a timer records how long a noise takes to travel between two microphones. The information recorded is:

   distance between microphones = 3 m
   time taken = 0.01 s

   Calculate the speed of sound.

2. When a wave on water is viewed from a beach there appears to be an upward and downward movement of the waves. What is actually transferred by the wave?

3. For the wave shown below calculate:

   (a) amplitude (b) wavelength
   (c) If 125 waves pass in 10 s what is the frequency?
   (d) Calculate the wave speed.

4. A water wave has a wavelength of 4 m and a frequency of 5 Hz. Calculate the wave speed.

5. A wave has a speed of 1500 m/s and a frequency of 2 MHz. Calculate the wavelength.

6. A wave has a speed of 600 m/s and a wavelength of 0.025 m. Calculate the frequency.

7. State the speed at which radio and TV waves are transmitted.

8. Explain what is meant by a transverse wave and give an example.

9. Explain why a sound wave is a longitudinal wave rather than a transverse one.

10. Put the following waves in the correct order going from high frequency to low frequency.

    gamma rays, TV ultraviolet, visible light, microwaves

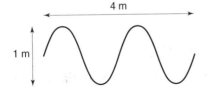

Figure 9.6 ▶

10 Reflection

At the end of this chapter you should be able to:

1 State what is meant by an optical fibre.

2 Describe one practical example of telecommunication which uses optical fibres.

3 State that electrical cables and optical fibres are used in some telecommunication systems.

4 State that light can be reflected.

5 Use the terms angle of incidence, angle of reflection and normal when a ray of light is reflected from a plane mirror.

6 State the principle of reversibility of a ray path.

7 Explain the action of curved reflectors on certain received signals.

8 Explain the action of curved reflectors on certain transmitted signals.

9 Describe an application of curved reflectors used in telecommunication.

10 Explain, with the aid of a diagram, what is meant by total internal reflection.

11 Explain, with the aid of a diagram, what is meant by the 'critical angle'.

12 Describe the principle of an optical fibre transmission system.

Law of reflection

A ray of light is shone from a ray box at a mirror. The angle of the ray is measured from a line drawn at right angles to the surface called a normal. This is the angle of incidence.

The angle of the reflected ray is measured. This is the angle of reflection. This can be repeated for other angles of incidence.

It is found that the angle of incidence equals the angle of reflection (figure 10.1).

Total internal reflection

A ray of light is shone into the semicircular glass block. As the angle of incidence is increased then at a certain angle no light will pass from

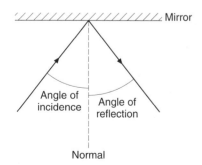

Figure 10.1 ▲
Reflection of light

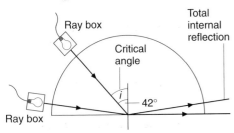

Figure 10.2 ▲
Total internal reflection of light

the semicircular block through the flat face of the block (figure 10.2). When no light passes from glass to air we have total internal reflection. The smallest angle of incidence at which total internal reflection occurs is called the critical angle.

The critical angle for glass is about 42°.

One use of total internal reflection is in fibre optics.

Optical fibres

Optical fibres are thin flexible glass threads about one eighth of a millimetre in diameter. Most of the glass thread is an outer layer of glass called cladding glass. In the middle of the thread is a different glass which is less than one hundredth of a millimetre in thickness. The fibres are made of extremely pure glass to cut down light loss and have a protective surface coating which reflects the light, keeping it inside the fibre.

An optical fibre is formed from glass so pure that a block 36 km thick would be as clear as an ordinary window pane. This means that we could see down to the bottom of the sea! Each fibre is no thicker than a strand of hair (figures 10.3 and 10.4).

Figure 10.3 ▶
Optical fibres

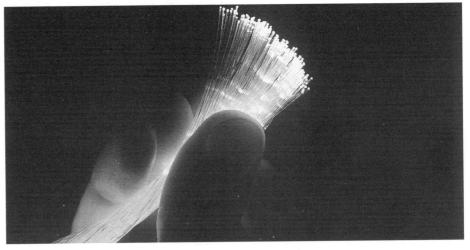

Figure 10.4 ▶
Light rays passing through an optical fibre

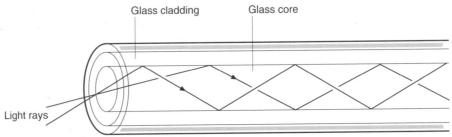

- Optical fibres are **lighter, carry more information** (up to 1000 telephone calls per fibre) and give better-quality communications than normal telephone wires.

- The signal that passes along the fibre is not electrical so it is less likely to be affected by other people's telephone calls or by other forms of electrical interference like mains hum.
- They are cheaper to make than copper since glass is mainly silica, which is cheaper than copper.
- The disadvantage is that it is more difficult to join fibres together than copper wires.

Many modern telecommunication systems use optical fibres instead of copper wires. One single hair-like fibre can carry all the information needed to bring telephone messages, cable TV, videotext and computer services into your home.

The fibres operate by the light being reflected down the fibre, since no light can leave the outside of the fibre due to the angle of incidence being greater than the critical angle. It has been estimated that one optical fibre cable could take all the telephone calls being used at once in the world. This could create very large bills.

Transmission and detection

To send speech information along glass fibres it is first necessary to change sound signals into suitable pulses of electrical energy. The microphone (transmitter) in a telephone hand-set and some microelectronics do this. These pulses of electricity control a small laser (a narrow, very powerful beam of light), which then produces pulses of light that are transmitted through the optical fibre (figure 10.5).

Figure 10.5 ▶
Use of fibre optics in a simple system

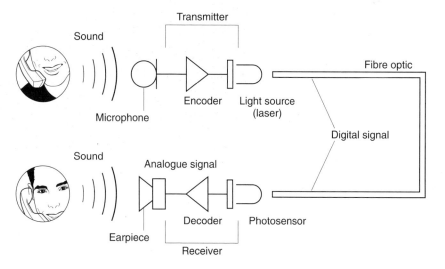

At the receiving end the light pulses are changed back into pulses of electricity by a small device called a photodiode. The electrical signal is fed to the earpiece (receiver), which then reproduces the original sound. In 1926, John Logie Baird invented a television system which used optical fibres, but it was not until some 40 years later that

Charles Kao and George Hockham suggested that optical fibres might replace copper wires for telecommunications. In 1977 the world's first optical fibre telephone system became operational in America and in 1978 they were used in a town in Manitoba, Canada to carry telephone, television, radio and computer information.

All of Britain's major cities are now linked by the major fibre link trunk system (figure 10.6). A transatlantic optical fibre link is now available.

Figure 10.6 ▶

Fibre optic installation, used to transmit telephone signals

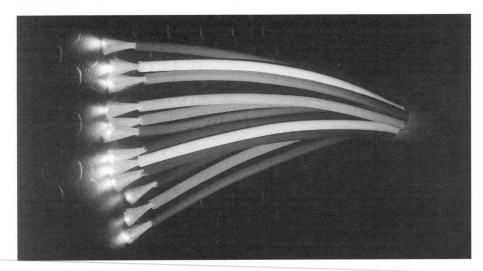

Modern optical fibres transmit light signals with very little signal loss and can be used over distances of about 100 km without amplification. With conventional copper cables there is so much loss (or attenuation) of the signal that repeater amplifiers have to be installed every 4 km.

Dish aerials and satellites

Nowadays we can send and receive both television and telephone signals from nearly any part of the world. To do this we need a transmitting aerial dish and a receiving dish. These then send and receive signals from a satellite orbiting the Earth.

Dish aerials

If a transmitting aerial is placed at the focus of a curved reflector (or dish), the reflected signal from the dish has the shape of a narrow beam (just like the light from a torch). This allows a strong (concentrated) signal to be sent in a particular direction from the transmitting dish aerial (figure 10.7).

Using dish aerials (curved reflectors)

When radio broadcasting began in 1920 the medium-frequency (MF) radio signals then used could travel about 1600 km. Soon after, high-frequency (HF) radio bands were discovered and these were used for worldwide communication.

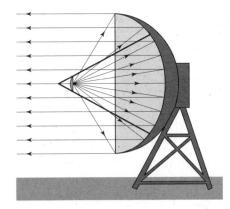

Figure 10.7 ▲

A transmitting dish aerial

Very high frequencies (VHF) soon followed, but these were unable to travel round the curvature of the Earth. Today VHF is used mainly for mobile communication, e.g. between aircraft and ground stations. As more and more information was transmitted the frequencies mentioned above became overcrowded and even higher frequencies had to be found. This led to the use of microwaves, whose frequency is about 10^9 Hz (1 GHz). Microwaves were useful since large amounts of information could be sent but little power was required. It was also found that they are easily focused using curved dish aerials (figure 10.8).

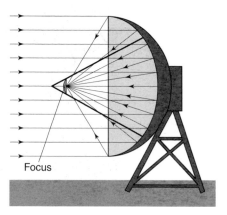

Focus

Receiving dishes gather in most of the signal and reflect it to one point called the focus. The receiving aerial is placed at the focus to receive the strongest signal.

Microwaves are unable to diffract (bend) round obstacles because of their small wavelength. This means that the transmitting and receiving dish aerials must be in line of sight and are often located on towers. Since microwaves have a fairly short range a series of repeater (relay) stations are needed every 40 km. The incoming microwaves are focused by the dish on to the receiving aerial. After being amplified (made bigger) they are passed to the aerial at the focus of the transmitting dish to produce a parallel narrow beam which is sent to the next repeater station. Using microwaves for round-the-world communication would require several hundred repeater (relay) stations at ground level. However, only three repeater stations in the sky, which are satellites, can cover all the earth if they are in the correct positions.

Satellites

A satellite has a very sensitive receiver as microwaves have to travel about 36 000 km to reach it. The signal, when received and focused, is amplified and transmitted back to Earth. Very large dish aerials on the Earth are required to pick up the weak signals coming from satellites and also to accurately send signals to them (figure 10.9). The dish to receive the signals is the same shape as the transmitting one but the signals are received in a parallel beam and the rays are then brought to a focus. The diagram is the same but the direction of the arrows is reversed.

Physics facts and key equations for reflection

- An optical fibre is a thin piece of glass along which light can be reflected.

- Modern television and telephone systems use electrical cables and optical fibres.

- When light is reflected from a plane mirror the angle of incidence = the angle of reflection.

- A curved reflector will bring the parallel rays of received signals to a focus, but on transmitted signals it will send them from the focus outwards as a parallel beam.

- Curved reflectors are used in satellite transmission and reception.

- Total internal reflection occurs when a ray of light does not leave a face of the glass but reflects from that face into the glass.

- The critical angle is the angle at which the ray inside the glass block just grazes the block on emerging.

- An optical fibre transmission system changes electrical signals into light, sends the light along a fibre system and then back to electrical signals at the receiving system.

End-of-Chapter Questions

1 (a) What is meant by an optical fibre?
  (b) Copy and complete the diagram to show the passage of light down an optical fibre.

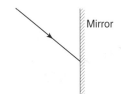

Figure 10.11 ▲

2 When light is shone onto a plane mirror it is reflected. Copy and complete the diagram to show what happens to the ray of light. Label your diagram to show the angle of incidence, angle of reflection and the normal.

Mirror

Figure 10.12 ▲

3 A local TV station is tranmitting radio waves from a curved transmitter as shown.

  (a) What happens to the rays as they leave the transmission point.
  (b) What is the correct name for the transmission point.

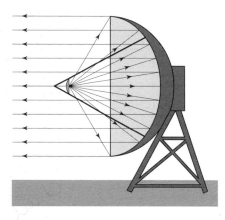

Figure 10.13 ▲

4 Why do satellite systems have two curved reflectors rather than just one. In your answer use a diagram to explain the function of each reflector.

5 Diamonds sparkle when cut and this is due to total internal reflection. What is meant by total internal reflection?

6 Modern communication systems use fibre optics to send information along them. At both the transmission and receiving ends we need to have electrical signals.

  (a) Explain the function of the fibre optic systems.
  (b) What is the main advantage of using fibre optics?

11 Refraction

At the end of this chapter you should be able to:

1 Describe the focusing of light on the retina of the eye.

2 State what is meant by refraction of light.

3 Draw diagrams to show the change of direction as light passes from air to glass and glass to air.

4 Use the correct terms angle of incidence, angle of refraction and normal.

5 Describe the shape of converging and diverging lenses.

6 Describe the effect of converging and diverging lenses on parallel rays of light.

7 Draw a ray diagram to show how a converging lens forms the image of an object placed at a distance of:

(a) more than two focal lengths
(b) between one and two focal lengths
(c) less than one focal length in front of the lens.

8 Carry out calculations involving the relationship between power and focal length of a lens.

9 State the meaning of long and short sight.

10 Explain the use of lenses to correct long sight and short sight.

Refraction of light

In a given material (called a medium) light travels in a straight line. When the light travels from one material to another it may bend or change direction as it enters the new material. When light enters a new material its speed changes. This effect is called **refraction**.

Rays of light can travel through various objects. The paths of the rays passing through and leaving the objects can be drawn as shown in figure 11.1. The dotted line drawn at right angles to the surface is called the normal.

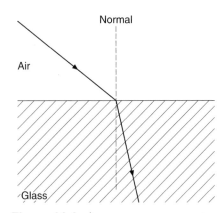

Figure 11.1 ▲
Refraction of light

- **Plane rectangular block**. When the incident ray travels parallel to the normal, there is no change in direction. When the incident ray is at an angle to a plane rectangular block, the ray coming from the block is parallel to the incident ray (figure 11.2).

Figure 11.2 ▶

Refraction of light through a rectangular block

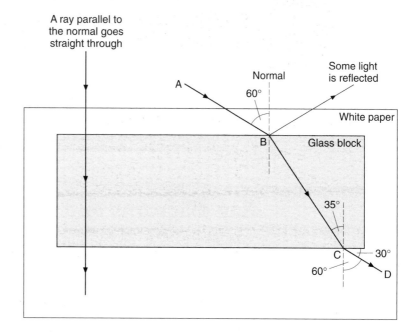

- **Triangular prism**. The ray bends towards the normal going into the prism and away coming out (figure 11.3).
- **Convex or converging lens**. The middle ray goes straight on and the outer rays bend and meet on the middle line at a point called the focus (figure 11.4). If the lens is thick, the same effect occurs but the focus is nearer the lens (figure 11.5).
- **Concave or diverging lens**. The rays spread out (figure 11.6).

Figure 11.3 ▼

Refraction of light through a triangular prism

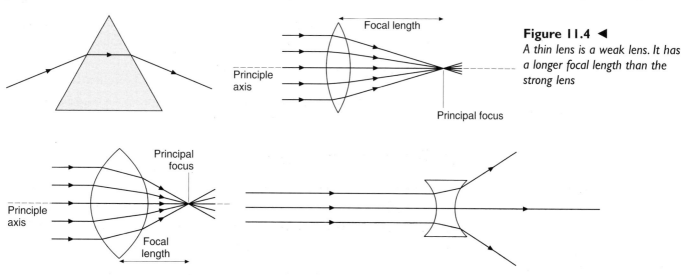

Figure 11.4 ◀

A thin lens is a weak lens. It has a longer focal length than the strong lens

Figure 11.5 ▲

A thick lens is a strong lens. It has a short focal length

Figure 11.6 ▲

A concave or diverging lens

Focal length

Some lenses bend light more than others, due to their thickness and the amount of curving of the lens. One way to indicate the amount of refraction is to measure the focal length of the lens. A convex or converging lens can make rays of light come together to a point after they have passed through the lens. The point where the rays meet is called the focus. The position of the focus depends on where the rays come from. When the rays come from a distant object which is so far away that the rays are parallel, the focus is closer to the lens. In this case, the focus is called the principal focus. The distance from the lens to the principal focus is called the **focal length** and is measured in metres.

Ray diagrams and lenses

An object such as a light source can be placed at different distances from a convex lens. The effect of placing the source at different distances has different effects on the final image. It is possible to know the effect by drawing an accurate diagram.

In doing these diagrams certain rules apply:

- The lens is considered thin enough to be a line and refraction only takes place at this line.
- Thousands of rays come from the object but it is possible to consider only the minimum number of rays to complete the diagram, which will be two rays.
- One ray is drawn from the top of the object parallel to the central line called the principal axis. This ray then passes through the principal focus.
- A second ray passes straight from the top of the object through the centre of the lens without changing direction.
- Where the rays intersect is the top of the object (figure 11.7).

Figure 11.7 ▶

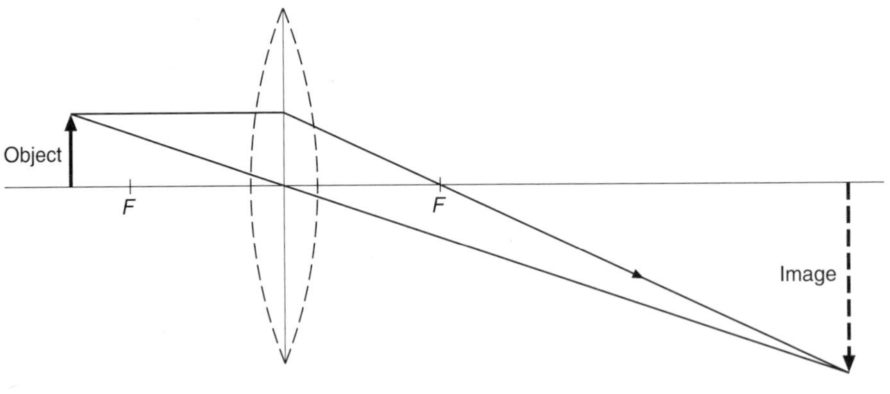

- The image can be described in three ways:

   (a) It is real or virtual. A real image can actually be seen on a screen. A virtual one is one that appears to come from that point. It is an illusion of the eye and brain and cannot be caught on a screen.

(b) It is magnified or diminished. If you measure the size of the image and compare it to the size of the object then you can easily find the change in size.

(c) Upright or inverted. The image will either be the same way up as the object or it will be upside down.

Three cases are shown in figure 11.8:

(a) Object more than two focal lengths: the image is real, inverted and diminished.

(b) Object between one and two focal lengths: the image is real, inverted and magnified.

(c) Object less than one focal length: here the rays do not meet unless they are traced back. The image is virtual, upright and magnified.

Figure 11.8 ▶

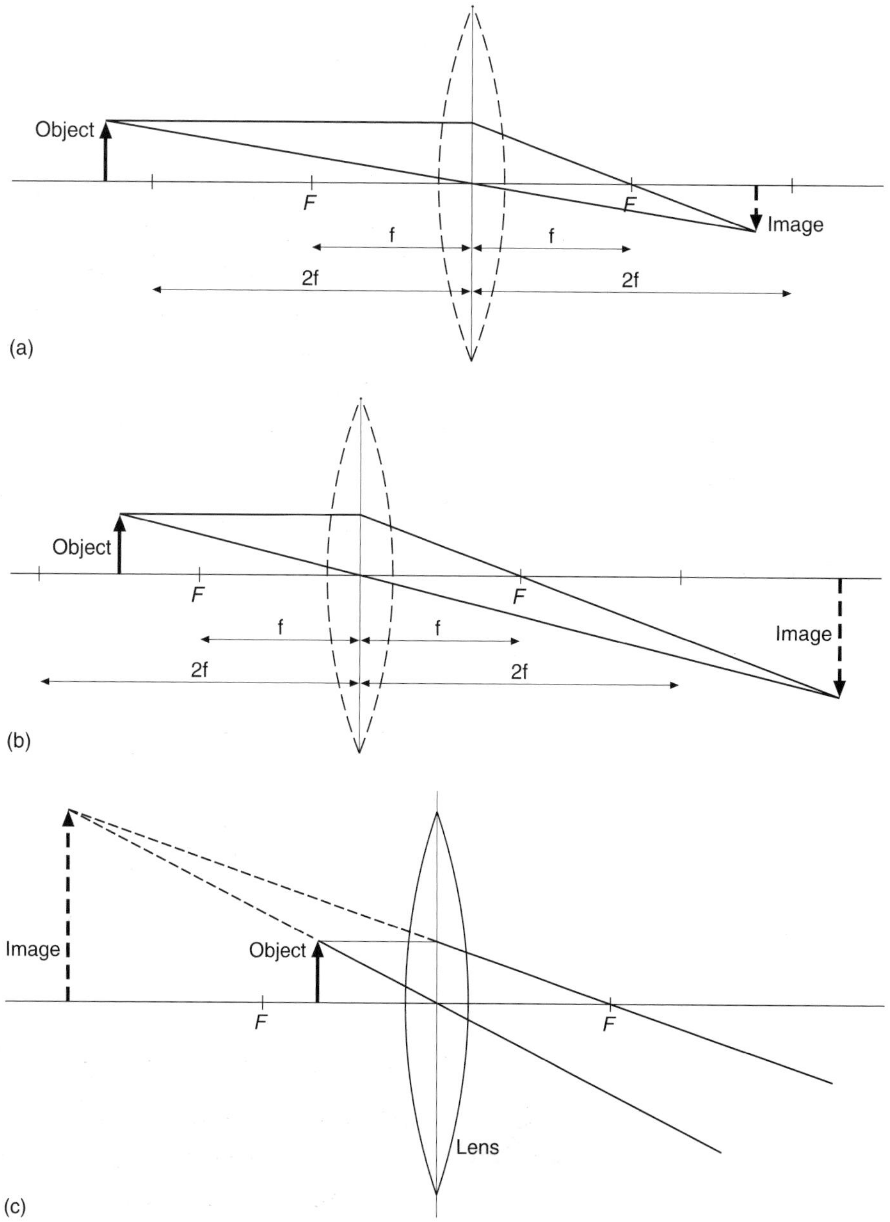

(a)

(b)

(c)

The last example is how a magnifying glass works. The eye thinks the image is magnified but the image is virtual since it cannot be caught on a screen.

The eye

Lenses are often used to correct eyesight defects. It is important to understand how the eye operates. An outline of the eye is shown in figure 11.9. The different parts of the eye and their functions are as follows:

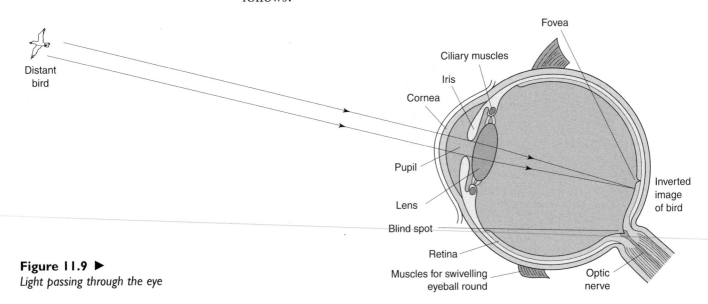

Figure 11.9 ▶
Light passing through the eye

- Light enters the front of the eye at the cornea. This is transparent and it is here that most of the refraction or bending of light occurs.
- The light enters a lens, which is a jelly-like substance, and more refraction occurs.
- The lens is held by fibres which act like muscles. These can change the shape of the lens from thick to thin.
- The light then passes through a gel-like substance which makes the light spread out.
- The light reaches the retina at the back of the eye.

Electrical signals pass along the nerve fibres to the brain. The part of the retina where the nerve fibres leave the retina contains no light-sensitive cells and is therefore a blind spot on the retina.

The amount of refraction which takes place at the cornea does not change. However, in order to focus on near and on distant objects, an adjustable lens is needed. This is provided by the eye's lens. This lens is held by muscle-like fibres in the ciliary body which can change the shape of the lens from thick to thin. The lens is thin and can focus on distant objects. To view near objects the muscles change the lens shape to thick. When light enters the eye, the image formed on the retina is upside down. The brain learns to turn this image the 'right way up'.

Power of a lens

People who have severe eyesight defects may need stronger (more powerful) lenses to correct their eyesight than those who have only slight defects. (A more powerful lens is one which causes more refraction.) An optician must therefore have a range of lenses to suit different individuals' needs. The powers of these lenses could be indicated by giving their focal lengths: the most powerful lenses having the shortest focal lengths. Another way to indicate the amount of refraction caused by a lens is to calculate its power from the equation:

$$\text{power} = \frac{1}{\text{focal length}}$$

where the focal length is measured in metres and the power is given in dioptres (D).

Converging (convex) lenses have positive powers (e.g. +10 D, +17 D).

Diverging (concave) lenses have negative powers (e.g. –10 D, –17 D).

Example

A convex lens has a focal length of 10 cm. Find the power of the lens.

Solution

focal length = 0.1 m

$$\text{power} = \frac{1}{\text{focal length}}$$

$$= \frac{1}{0.1}$$

$$= 10 \text{ D}$$

Example

A lens has power of –2 D. Calculate its focal length.

Solution

Power = –2 D, which tells us that this a concave lens since there is a negative sign.

$$\text{Power} = \frac{1}{\text{focal length}}$$

$$2 = \frac{1}{\text{focal length}}$$

$$\text{focal length} = \frac{1}{2}$$

$$= 0.5 \text{ m}$$

Long sight

A long-sighted person can see far-away objects clearly. Objects quite close to the eye appear blurred. The eye lens is bringing the rays to a focus beyond the retina. This may happen if the person's eyeball is shorter than normal from front to back. It may also be caused by ciliary muscles which cannot relax for the lens to be made fat enough.

A converging (convex) lens corrects this fault since it will increase the bending of the light rays before they enter the eye lens and so the light will be focused on the retina. The object is thus seen clearly (figure 11.10).

Figure 11.10 ▶
(a) A long-sighted eye cannot see near objects clearly. (b) A converging lens corrects long sight.

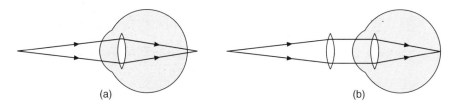

(a)   (b)

Short sight

A short-sighted person finds that distant objects are blurred but near objects are in focus. The eye lens is bringing light to a focus in front of the retina, which may happen due to the lens having a large curvature. The muscles cannot make the lens thin enough.

A diverging (concave) lens corrects this fault, since it will spread the light out more before it enters the eye lens, and so the light will be focused on the retina (figure 11.11).

Figure 11.11 ▶
(a) A short-sighted eye cannot see distant objects clearly. (b) A diverging lens corrects short sight.

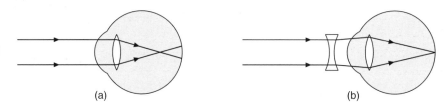

(a)   (b)

Did you know?

The first spectacle lenses were developed around about 1270 but there is a legend that the emperor Nero had an emerald cut as a lens to enable him to see more clearly.

Contact lenses were first suggested by Leonardo da Vinci in 1518, who noticed that he could see more clearly when he opened his eye under a bowl of water. Different lenses have been developed that allow oxygen to pass into the eye, which helps to prevent some eye diseases. There are lenses which can be thrown away after use each day and also ones which have a varying degree of curvature (figure 11.12).

Figure 11.12 ▶
Contact lenses

Physics facts and key equations for refraction

- Refraction of light occurs when light travels from one substance to another.

- When light travels from air to glass the ray bends towards the normal and the angle of incidence is greater than the angle of refraction.

- A converging lens brings parallel rays to a focus but a diverging lens spreads the rays out.

- For a converging lens, the image of an object placed at a distance of:

  (a) more than two focal lengths is real, diminished and inverted
  (b) between one and two focal lengths is real, magnified and inverted

  (c) less than one focal length is virtual, magnified and upright.

- The power of a lens in dioptres is calculated as

$$\text{Power} = \frac{1}{\text{focal length}}$$

  where the focal length is in metres.

- Long sight is when distant objects are seen clearly but near objects are blurred. A converging lens will correct the problem.

- Short sight is when distant objects are blurred but near objects are clear. A diverging lens will correct the problem.

End-of-Chapter Questions

1 A ray of light passes from air to glass as shown in Figure 11.13.

  (a) What is the name for this effect?

  (b) Copy the diagram and label the angle of incidence and the angle of refraction.

  (c) What can you state about the size of the angle of incidence compared to the angle of refraction?

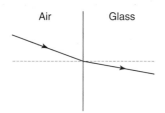

Figure 11.13 ▲

2 In the diagram shown below the object is a plant which is viewed at a distance of 25 cm through a convex lens of focal length 10 cm.

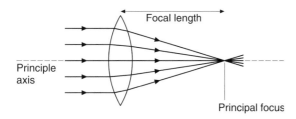

Figure 11.14 ▲

  (a) Draw a suitable diagram to scale to show the final image.

  (b) By measuring the height of the final image calculate the change in size of the image compared to the object.

3 A postage stamp is viewed using a magnifying glass. As the glass is moved closer to the stamp the image changes from upside down to upright.

  (a) At what lens distance will this effect begin to occur?

  (b) The lens has a focal length of 5 cm and the stamp is placed 3 cm from the centre of the lens.

  Show by a suitable diagram what happens to the rays of light and calculate the magnification.

4 A slide projector is used to show some optical effects. The slide is placed at 15 cm from the centre of a lens of focal length 10 cm.

  Draw a diagram to show the final image.

5 A lens prescription states that a correcting lens of + 2.5 D is required.

  (a) What type of eye defect is the person suffering? Explain your answer.

  (b) Calculate the focal length of the lens.

6 A person can read this book without the aid of glasses but needs spectacles to see distant objects.

  (a) Name the sight defect.

  (b) What type of lens is needed to correct the defect?

  (c) The focal length of this lens is 60 cm. Calculate the power of the lens needed.

Exam Questions

1 A radio controlled toy operates on a frequency of 27 MHz.

(a) (i) Explain what is meant by 'frequency of a wave'.

   (ii) Calculate the wavelength of this transmitter.

(b) Modern communication systems use fibre optics as a method of sending signals.

   (i) What is meant by a fibre optic sytem?

   (ii) An optical fibre can normally send a signal a distance of 40 km before amplification is required.

Calculate the time rquired to send this signal down the fibre. The speed of light in the fibre is 2×10^8 m/s.

2 During a test of lenses, a small lamp of height 1 cm is placed at a distance of 30 cm from a convex lens of focal length 20 cm.

(a) (i) By using a suitable scale diagram find where the image is located.

   (ii) State whether the image is real or virtual and whether it is magnified or diminished.

(b) This same lens can be used by someone with an eye defect.

   (i) What eye defect would be corrected by using a convex lens?

   (ii) Calculate the power of this lens.

Radioactivity

  **Ionising radiation**

After reading this chapter you should be able to:

1 Describe a simple model of the atom which includes protons, neutrons and electrons.

2 State that radiation energy may be absorbed in the medium through which it passes.

3 State the range and absorption of alpha, beta and gamma radiation.

4 Explain what is meant by an alpha particle, beta particle and gamma radiation.

5 Explain the term ionisation.

6 State that alpha particles produce much greater ionisation density than beta particles or gamma rays.

7 Describe how one of the effects of radiation is used in a detector of radiation.

8 State that radiation can kill living cells or change the nature of living cells.

9 Describe one medical use of radiation based on the fact that radiation can destroy cells.

10 Describe one use of radiation based on the fact that radiation is easy to detect.

Atoms are the smallest possible particles of the elements which make up everything around us. All atoms of the one element are identical to one another, but they are different from all other elements. This is because they are made up from different combinations of electrons, protons and neutrons. These are the three types of particles which make up an atom.

All atoms have a tiny central nucleus, which has a positive charge. We can imagine the negatively charged electrons to be circling around this nucleus. The nucleus contains the positive protons and the neutrons, which are uncharged (figure 12.1).

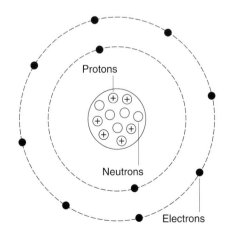

Figure 12.1 ▲
Diagram of an atom

Different types of radiation

There are three types of radiation:

1 Alpha (α) particles.

2 Beta (β) particles.

3 Gamma (γ) rays.

These can be identified by what happens as they reach different materials.

Alpha radiation can be completely stopped by a few sheets of paper.

Beta radiation can be absorbed by a sheet of aluminium.

Gamma radiation is reduced and may be completely absorbed by materials such as concrete, lead and other dense materials. The thicker the material the more radiation that will be absorbed. This use will be discussed later as a form of protection.

When the alpha or beta or gamma radiation passes through a material it loses energy by colliding with the atoms of the material. Eventually the radiations lose so much energy that they cannot get through (penetrate) the material and so are absorbed.

The three different types of radiation and their natures are shown in the following table.

Type of radiation	Symbol	What is this radiation	Charge and absorption
Alpha	α	Helium nucleus, i.e. two protons and two neutrons	2+ Absorbed by paper or about 5 cm of air
Beta	β	Fast-moving electron	1– Absorbed by aluminium
Gamma	γ	Short-wavelength electromagnetic radiation	Uncharged Absorbed by lead

Ionisation

If an electron is added or removed from an atom, what is left is called an **ion**. The process is called ionisation. The removal of an electron creates a positive ion and if an electron is added then a negative ion is formed.

The process of ionisation by an alpha particle is shown in figure 12.2. In (a) the alpha particle is approaching the neutral atom and in (b) it has passed by having created an ion pair. This means that the alpha particle has caused the atom to lose an electron and form an ion.

Figure 12.2 ▶
(a) Neutral carbon atom. (b) Ionisation of carbon atom

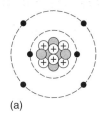 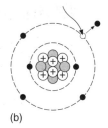

(a)   (b)

Since an alpha particle moves much slower than beta or gamma radiation it will cause much more ionisation. The amount of ionisation produced in a certain volume is called the **ionisation density**.

Effect on the human body

- Alpha radiation will produce ionisation in a short distance of body tissue. This type of radiation outside the body is absorbed by the skin and little damage will occur. Swallowing the radiation will produce large amounts of ionisation and would be dangerous.
- Beta radiation will be absorbed in about 1 cm of tissue and any beta radiation outside the body will cause damage to that tissue but a small amount can penetrate the body. If the radioactive source is put into the body then internal organs can be damaged.
- Gamma rays and X-rays pass through the body and can damage tissue whether the source is inside or outside of the body.

Sterilisation

Radiation can be used to kill cells, and can also be used to kill bacteria or germs. Previously, medical instruments such as glass syringes had to be sterilised by heat or chemicals. Now cheap, plastic, throw-away syringes can be used. They are pre-packed and then irradiated by an intense gamma ray source. This kills any bacteria but does not make the syringe radioactive.

Detecting radiation

Photographic fogging

Photographic film has a thin layer of silver-based chemical on the surface of the plastic or paper. Normally, this silver salt is affected by light falling on it – wherever it lands, it changes the chemical and blackens or fogs the film surface.

Alpha, beta or gamma radiation has a similar effect on this photographic emulsion, and so photographic film can be used to detect them. In fact, radioactive substances were first discovered, by accident, when Henri Becquerel, a French physicist, left some uranium rocks near photographic paper. He discovered that the paper had been blackened, This work led Pierre and Marie Curie to investigate these effects and gave us the term radioactivity.

Workers who use radioactive materials, particularly health workers in hospitals, must wear film badges throughout their working day. This enables a check to be made on the amount of radiation to which they have been exposed. When the film is developed, the amount of fogging gives a measure of the radiation exposure (figure 12.3). Different windows are used to measure the amounts of the different types of radiation.

A plastic window will absorb different energies of beta rays. Metal windows absorb different energies of gamma and X-rays. Aluminium will absorb low-energy X-rays. The other metals will absorb the high-energy X-rays.

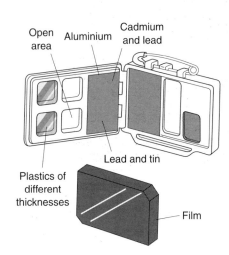

Figure 12.3 ▲
A film badge, for indicating exposure of a worker to radioactivity

Scintillations

Some substances such as zinc sulphide are fluorescent. This means that they absorb radiation and give out energy again as a tiny burst of light. These flashes of light are called scintillations, and they may be observed by the naked eye or counted by a light detector and an electronic circuit. These scintillation counters are used in many modern instruments, including the gamma camera.

Using radiation in medicine

Treating cancer: radiation therapy

Radiotherapy is the treatment of cancers by radiation. Cancers are growths of cells which are out of control. Cancerous tumours can be treated by chemotherapy (very powerful drugs with side effects), surgery or radiation. The choice of treatment depends on the size and position of the tumour. Very often radiotherapy is used after surgery to destroy any remaining cancerous cells. The object of the radiation treatment is to cause damage to the cancer cells, which then stop reproducing. The tumour then shrinks.

Unfortunately, healthy cells can also be damaged by radiation. The amount of radiation has to be very accurately calculated so that sufficient damage is done to cancer cells without overdoing the damage to other cells. The radiation must be aimed very accurately at the tumour. This can be done using a simulator. A series of X-ray photographs are taken at different angles and a computer can build up a picture of the tumour and measure the amount of radiation to be given. Some localised tumours (e.g., a bone tumour) can be treated by irradiation with high-energy X-rays or gamma rays (figure 12.4).

Figure 12.4 ▶
Radiotherapy

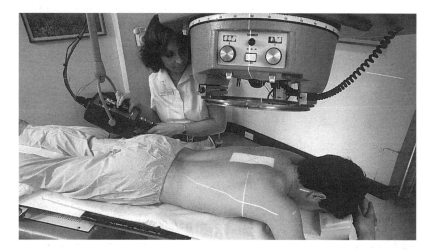

The gamma rays are emitted from a cobalt-60 source – a radioactive form of cobalt. The cobalt source is kept within a thick, heavy metal container. This has a slit in it to allow a narrow beam of gamma rays to emerge.

The X-rays are generated by a linear accelerator. This machine fires high-energy electrons at a metal target and when the electrons strike the target X-rays are produced. The X-rays are shaped into a narrow beam by movable metal shutters.

With either technique, the apparatus is arranged so that it can rotate around the couch on which the patient lies. This allows the patient to receive radiation from different directions. Rotating the source of radiation means that the tumour is always receiving radiation but the healthy tissue only receives a fraction of this radiation. Treatments are given as a series of small doses because tumour cells are killed more easily when they are dividing, and not all cells divide at the same time. This reduces the side effects, such as sickness.

The gamma camera

It is important for doctors and scientists to be able to study internal organs without surgery. A **radiopharmaceutical** is used which can act as a tracer. The advantage of the radiopharmaceutical is that the radiation can be detected outside the body.

The radiopharmaceutical has two parts (figure 12.5):

Figure 12.5 ▶
Formation of a radiopharmaceutical

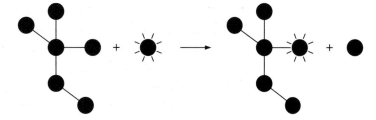

- A drug which is chosen for the particular organ that is being studied. Different organs have different drugs.
- A radioactive substance which is a gamma emitter.

Gamma is chosen since alpha or beta would be absorbed by tissue and would not be detected outside the body.

The combined substance is then injected into the patient.

To detect the radiation a **gamma camera** is used. This uses special crystals which emit flashes of light when the radiation reaches them. Photomultiplier tubes can change the light energy into electrical energy. The signals can then be displayed on a screen (figure 12.6).

Figure 12.6 ▶
Gamma camera

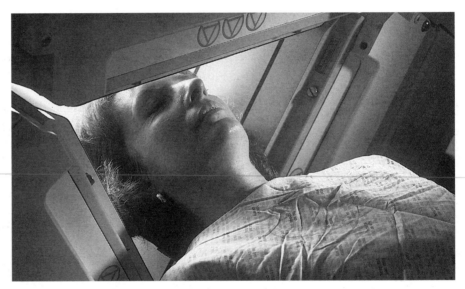

There are two types of studies:

- A static study is where there is a time delay between injecting the radioactive material and the build-up of radiation in the organ. This occurs in bone, lung or brain scans.
- A dynamic study is where the amount of radioactivity in an organ is measured as time increases. This occurs in the examination of the operation of the kidneys. This is called a renogram. The latter technique examines the working of the kidneys. The radioactive material reaches the kidneys due to the drug given to the patient. The radioactive material is taken out of the bloodstream by the kidneys. Within a few minutes of the drug being injected the radiation is concentrated in the kidneys. After 10–15 minutes, almost all the radiation should be in the bladder. The gamma camera takes readings every few seconds for 20 minutes. The computer adds up the radioactivity in each kidney. This can be shown as a graph of activity against time (figure 12.7). Both kidneys take up the radioactivity. The radioactivity does not decrease as rapidly from the left kidney as it does from the right. The left kidney is not working correctly.

Figure 12.7 ▶

A renogram. The renogram shows the time-activity for a kidney. It can be analysed to produce crucial information. In this example the left kidney is completely normal.

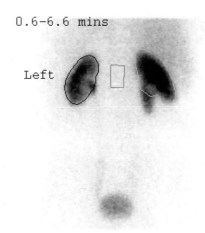

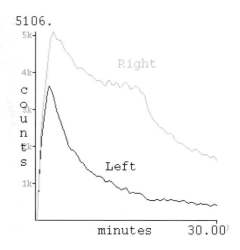

Physics facts and key equations for ionising radiation

- The atom includes protons (+), neutrons and electrons (–).

- Radiation energy may be absorbed in a substance.

- Radiation can kill or change the nature of living cells.

- Alpha radiation is absorbed by paper and will travel about 5 cm in air. Beta is absorbed by aluminium and gamma radiation is reduced by lead or concrete.

- An alpha particle is a helium nucleus; a beta particle is a fast-moving electron and gamma radiation is a short-wave electromagnetic radiation.

- Ionisation is the gain or loss of an electron to produce a charged particle.

- Alpha particles produce much greater ionisation density than beta particles or gamma rays.

- Radiation is used in a detector of radiation.

- Radiation can destroy cells and is used in cancer treatment.

- Gamma cameras use radiation to examine the organs of the body.

End-of-Chapter Questions

1 (a) State the three particles present in an atom and the charge on each one.
(b) Where is each particle located in the atom?

2 Alpha particles are emitted from a source.

(a) (i) What is an alpha particle?
(ii) What is the thinnest material which will absorb alpha particles?
(b) Ionisation occurs with alpha particles. What is meant by ionisation?
(c) Why is it more dangerous to have an alpha source inside your body than a beta or gamma source?

3 Gamma radiation is used to detect if an organ in a patient is working correctly.

(a) Give an example of this and state which instrument is used to detect the radiation.
(b) Why is gamma radiation used rather than beta or alpha radiation?

4 (a) What is a beta particle?
(b) What is the thinnest material which will absorb beta radiation?

5 Radiation can kill or cure in certain cases. Describe a use of radiation to cure a patient.

13 Dosimetry

After reading this chapter you should be able to:

1 State that the activity of a radioactive source is measured in becquerels, where one becquerel is one decay per second.

2 State that the absorbed dose *D* is the energy absorbed per unit mass of the absorbing material.

3 State that the gray is the unit of absorbed dose and that one gray is one joule per kilogram.

4 State that the risk of biological harm from an exposure to radiation depends on

   (a) the absorbed dose
   (b) the kind of radiation
   (c) the body organs or tissue.

5 State that a quality factor *Q* is given to each kind of radiation as a measure of its biological effect.

6 State that the equivalent dose *H* is the product of *D* and *Q* and is measured in sieverts, Sv.

7 Carry out calculations using $H = DQ$.

8 Describe factors affecting the background radiation level.

Activity

All ionising radiation, that is, alpha, beta, gamma and X-rays, can cause damage to the cells of the body. It should be stressed that there is no minimum amount of radiation which is safe. The aim in physics is to measure the radiation and to estimate the risk when we are exposed to radiation. Some of this radiation occurs naturally but we still need an estimate of risk.

The becquerel

This unit measures the number of atoms which disintegrate or break up each second. If 50 atoms break up each second then the activity of the source is 50 becquerels, abbreviated as Bq.

In practice, particularly in medical treatments, this is too small and larger units such as kBq and more often MBq (mega) are used. However, one gram of plutonium used in nuclear reactions has an activity of 2000 MBq

Absorbed dose

When radiation reaches the body or tissue it is absorbed. This is called the absorbed dose (*D*) and is equal to one joule of energy absorbed by one kilogram of tissue.

This can be written as $D = \dfrac{E}{m}$

The unit is called the gray (Gy) and typical values are shown in the following table for a number of medical procedures.

Radiation treatment	Absorbed dose (Gy)
Chest X-ray	0.00015
CT scan	0.05
Gamma rays which would just produce reddening of the skin	3.0
Dose which if given to whole body in a short period would prove fatal in half the cases	5.0
Typical dose to a tumour over a six-week period	60.0

This has to be considered as a rough method of gauging response to radiation. For a typical person exposed to 10 Gy of radiation the effect could be fatal, yet if received as heat energy your body temperature would only rise by a few thousandths of a degree.

In addition, radiation received by the body has a different effect depending on the type of radiation and the organ which receives it.

Equivalent dose

Tissue can receive the same amount of radiation but from different sources. In each case the effect on the tissue will be different.

To take account of the different types of radiation a **quality factor**, *Q*, is used. The quality factor for X-rays, gamma and beta radiation is 1 but for alpha radiation it is 20. This is because of the high ionisation density of alpha radiation. When the quality factor is taken into account a quantity called the equivalent dose is measured. The unit of equivalent dose is sieverts (Sv) and is given the symbol *H*:

$$H = DQ$$

For each part of the body there will be different effects and the calculation of the absorbed equivalent dose takes into account the different types of tissue. For example, the lungs have a weighting of 0.2 and the bladder of 0.5. The use of these units can then allow us to estimate the risk of fatal cancer, which is defined as 5% per sievert. This means that if 100 people were exposed to an equivalent dose of

1 Sv to the whole body then five would develop a fatal cancer. In practice such exposures are not possible except in extreme accidents.

In research into radiation exposure, no definite health effects have been reported up to 100 mSv (0.1 Sv).

At 10 Sv, radiation sickness, skin burns and an increase in cancer can result.

Typical values of equivalent dose are shown below.

Investigation	Equivalent dose (mSv)
Chest X-ray	0.1
Spine X-ray	2.0
Stomach X-ray	4.0
CT scan	1 to 3.5
Bone scan	2.0
Annual exposure of aircraft crew	2.0
Renogram	2.0
Astronaut in space for one month	15

Example

A 50 kg person is exposed to radiation of energy 0.25 J. The quality factor for the radiation is 20.

(a) Calculate the absorbed dose for this radiation.
(b) What is the equivalent dose?

Solution

(a) $D = \dfrac{E}{m}$

$D = \dfrac{0.25}{50}$

$= 0.005 \text{ Gy}$

(b) $H = DQ$

$= 0.005 \times 20$

$= 0.1 \text{ Sv} = 100 \text{ mSv}$

It is important to put these figures in context. One mSv is about 100 times the radiation you experience when you travel by aircraft on holiday. But if you are part of the aircrew, then you will experience larger amounts due to the amount of travel. There are regulations about total flying times which take into account exposure to radiation.

It is also about half the annual dose of radiation which you will receive anyway from natural radiation.

Background radiation

There is radiation all around us. This is known as background radiation, and it is almost all from natural sources. Tables show the typical equivalent dose we get every year from background radiation and other sources. They are average figures, and vary a lot depending on our job and where we live.

Source	Annual dose (µSv)
Natural sources	
Radon and thoron gas from rocks and soil	800
Gamma rays from ground	400
Carbon and potassium in body	370
Cosmic rays at ground level	300
Total = 1870 µSv, that is, 1.87 mSv.	
Man-made sources	
Medical uses: X-rays etc.	250
Fallout from weapons testing	10
Job (average)	5
Nuclear industry (e.g. waste)	2
Others (TV, aeroplane trips, etc.)	11
Total = 278 µSv, that is, 0.278 mSv.	

Radon gas

This is radiation from granite rock structures and is higher in some parts of the country than in others. The rocks contain radium which decays to produce radon gas. The design of modern houses actually prevents its dispersal easily since double glazing and the lack of chimneys means the gas cannot escape. The gas can be prevented from reaching the rooms by sealing the ground and making underfloor ventilation. In the USA the Environmental Protection Agency calls it the second largest cause of lung cancer.

Gamma rays

These are mainly emitted in substances like uranium and thorium which are contained in building materials.

Carbon and potassium

Some radioactivity occurs in air, food and water. The largest amount is from potassium-40. This amount of radiation is controlled by our muscular systems. The average activity in our body is 4000 Bq.

Cosmic rays

This is radiation which comes from outer space. The Earth's atmosphere reduces our exposure. However, they can penetrate buildings and enter our bodies.

Medical uses

The largest man-made radiation dose comes from medical sources. This covers X-rays, scans and cancer treatment. This figure is an average figure since unless you are receiving treatment you may receive only small amounts due to dental X-rays, for example.

Fallout

This is the name given to the radiation from nuclear weapons testing. Although there are very few tests taking place today, the radiation still exists from previous work.

Industry

Some jobs will involve the use of radiation. In heavy engineering, radiation is used to test welding joints. This technique is known as non-destructive testing. Miners also receive greater radiation due to thoron and radon gases, and aircrews will also receive larger amounts.

Nuclear industry

Nuclear reactors will discharge small amounts of radioactivity into the atmosphere.

Others

Any TV monitor or computer monitor will emit very small amounts of radiation. All smoke detectors use a radioactive source.

The different amounts are shown in figure 13.1. You can see that natural radiation is by far the biggest influence on us.

The total annual equivalent dose in the UK averages about 2000 μSv, or about 2 mSv. But there is a big variation from person to person.

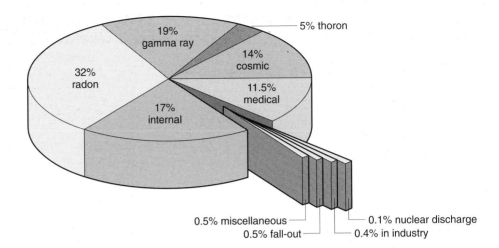

Figure 13.1 ▶
Distribution of radiation

Physics facts and key equations for dosimetry

- The activity of a radioactive source is measured in becquerels, where one becquerel is one decay per second.

- The absorbed dose D is the energy absorbed per unit mass of the absorbing material and is measured in grays, which are joules per kilogram.

- The risk of biological harm from radiation depends on:

  (a) the absorbed dose

  (b) the kind of the radiation

  (c) the body organs or tissue exposed to the radiation.

- Quality factor Q is given to each kind of radiation as a measure of its biological effect.

- Equivalent dose $H = DQ$ and is measured in sieverts (Sv).

- There are several sources of background radiation, which can be split into natural and man-made.

1 What is meant by the activity of a radioactive source and state the units in which it is measured.

2 A radioactive source has an activity of 140 Bq which is absorbed directly by a person of mass 50 kg.

Calculate the absorbed dose.

3 State the three factors which affect the risk of biological harm from radiation.

4 State the equation for equivalent dose and the units in which it is measured.

5 A radioactive source emits radiation which is absorbed by a person of mass 50 kg. The absorbed dose is 0.05 Gy. If the quality factor for this radiation is 10, calculate the equivalent dose for this person.

6 Copy and complete the table.

Absorbed Dose/Gy	Quality factor	Equivalent dose/Sv
0.001	1	
0.003	20	
	20	0.002
	1	0.01
0.005		0.01
0.002		0.04

7 State two factors which affect background radiation.

14 Half-life and safety

At the end of this chapter you should be able to:

1 State that the activity of a radioactive source decreases with time.

2 State the meaning of the term 'half-life'.

3 Describe the principles of a method for measuring the half-life of a radioactive source.

4 Carry out calculations to find the half-life of a radioactive element from appropriate data.

5 Describe the safety precautions necessary when handling radioactive substances.

6 State that the equivalent dose is reduced by shielding, by limiting the time of exposure or by increasing the distance from a source.

7 Identify the radioactive hazard sign and state where it should be displayed.

Half-life

When a radioactive substance disintegrates, the activity (number of disintegrations or source count rate) depends only on the number of radioactive nuclei present, i.e. double the number, double the activity. The activity of all radioactive materials will decrease with time.

The half-life of a radioactive substance is the time taken for half the radioactive nuclei to disintegrate i.e. the time taken for the activity to fall by one half.

Half-life is measured in units of time. This could be in seconds, minutes, days or years.

Typical half lives are

- Uranium-238: 4.5×10^9 years
- Radium-226: 1600 years
- Cobalt-60: 5.3 years
- Sodium-24: 15 hours
- Copper-66: 5.2 minutes

Total count rate and background count rate

In any radioactive experiment a count rate will be obtained even if no radioactive source is present. This is due to background radiation. The background count rate could be found by measuring the number of background counts in a known time and working out an average count rate (counts per second or counts per minute) from this. Alternatively, when carrying out your experiment you could measure the total count rate with your measuring device, where:

total count rate = count rate from radioactive source
+ count rate due to background

Measuring half life

To measure half-life of a radioactive source in a laboratory you will need a source and a detector/counter. This is shown in figure 14.1. The detector is usually a Geiger counter.

Figure 14.1 ▶
Geiger counter

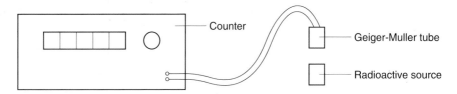

You would take the background count rate several times without the source being present.

Take an average of your background count and calculate the background count rate.

Activate the radioactive source. This is done by shaking the source since in laboratories it is a weak combination of two liquids. This source is then placed over the detector at a fixed distance and the count rate taken. Generally the counter will record for say 10 seconds, hold the reading for 2 seconds and then repeat this process. If you allow this to continue you will obtain a set of readings of count rate against time. Before you analyse the readings you will need to subtract the background count rate to obtain the true count rate due to the source alone. You should then either calculate the half-life or draw a graph of count rate against time, which will allow you to see the pattern more easily. In carrying out this experiment you should comply with the safety rules concerning radioactive sources, which are detailed later.

After one half-life ($t_{1/2}$), the activity and the measured count rate from the source drop to half the initial value. After a second half-life, the count rate halves again – it is now one quarter of its original value. Three half-lives will see the activity and count rate reduce to one eighth, and so on.

A graph of count rate against time is shown in figure 14.2.

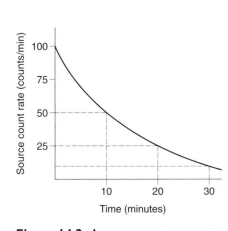

Figure 14.2 ▲
A graph of count rate from a radioactive source against time

Source count rate corrected for background	Time in minutes
100	0
50	10
25	20
12	30
7	40

Average half life is 10 minutes.

Example

A radioactive source gives an initial activity of 3200 Bq. After 150 minutes the activity is found to be 100 Bq. Calculate the half-life of the source.

Solution

After 1 half-life the activity is 1600 Bq.

After 2 half-lives the activity is 800 Bq.

After 3 half-lives the activity is 400 Bq.

After 4 half-lives activity is 200 Bq.

After 5 half-lives activity is 100 Bq.

5 half-lives = 150 minutes.

1 half-life = 30 minutes.

Example

A radioactive source has a half-life of 4 days. At the start of an experiment the total activity recorded is 990 counts per minute. Find the total recorded activity after 20 days, if the background count rate is 30 counts per minute.

Note:

1 Total count rate = count rate of source + background count rate.

2 The background rate is constant and does not decrease with time.

Solution

count rate of source $= 990 - 30$

$= 960$ counts per minute

number of half lives $= \dfrac{20}{4}$

$= 5$

After 1 half-life the count rate is 480 counts per minute.

After 2 half-lives the count rate is 240 counts per minute.

After 3 half-lives the count rate is 120 counts per minute.

After 4 half-lives the count rate is 60 counts per minute.

After 5 half-lives the count rate is 30 counts per minute.

Total recorded count rate is 30 + 30 = 60 counts per minute.

There are people who, as a result of their work, are exposed to ionising radiations. These people include medical workers using X-rays in hospitals, dentists and vets, research workers using radiation sources in their experiments and people who work in the nuclear power industry.

Protection using radiation

Just after the discovery of radioactivity in 1896, radiation was thought to be good for health. Radium was advertised as a fertiliser for crops. Radium, now known as a major hazard, was sold in inhalers and an article in 1925 claimed that it had cured 27 illnesses from anaemia to sciatica. In addition during the 1920s the ladies who painted the luminescent faces on watches later developed cancer of the jawbone due to the sucking of their brushes. This led to changes in attitudes about radiation and steps to reduce exposure.

There are three methods by which radiation exposure can be reduced:

1 *Shielding a source of radiation with an appropriate thickness of absorber.* A radiographer wears a lead-lined apron. Radioactive sources are contained in lead containers.

2 *Limiting the time.* Sources should be moved and used as quickly as possible to reduce the radiation present. In the early days of radiation some workers changed photographic plates for X-rays with the beam still on. This resulted in radiation burns to the skin of the hand.

3 *Distance from source.* The further you are from the source the less radiation you will receive. In fact if you double the distance you will only receive a quarter of the radiation.

Safety with radioactivity

1 Always use forceps or a lifting tool to remove a source (figure 14.3). Never use bare hands.

2 Arrange a source so that its radiation window points away from the body.

3 Never bring a source close to your eyes for examination.

Figure 14.3 ▶
Lifting a radioactive source with forceps

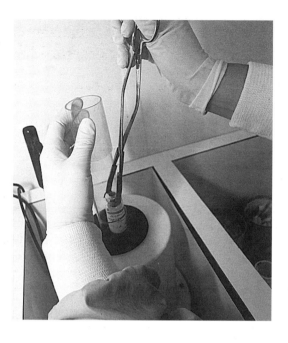

4 When in use, a source must always be attended by an authorised person and it must be returned to a locked and labelled store in its special shielded box immediately after use.

5 After any experiment with radioactive materials, wash your hands thoroughly before you eat.

6 In the UK, students under 16 years old may not normally handle radioactive sources.

The symbol for radiation sources being stored is shown in figure 14.4. It must be displayed where radioactive materials are being used or stored, perhaps in a school or college. You will see it on doors in hospital areas and on boxes of radioactive sources which are being transported. It is an international symbol.

Figure 14.4 ▲
Radiation hazard symbol

 **Physics facts and key equations for half-line and safety**

- The activity of a radioactive source decreases with time.

- The half-life of a radioactive substance is the time taken for half the radioactive nuclei to disintegrate.

- To measure the half-life of a radioactive source you will need a radioactive source, a detector and a timer. You will need to measure the activity of the source against time.

- Describe the safety precautions necessary when handling radioactive substances.

- The equivalent dose is reduced by shielding, by limiting the time of exposure or by increasing the distance from a source.

- The radioactive hazard sign is an international one and it should be displayed where sources are stored or are in daily use.

End-of-Chapter Questions

1 What is meant by 'a half-life of a radioactive source is 6 hours'?

2 A radioactive source has an activity of 640 kBq. The half-life of the source is 8 days.

What will be the activity after 40 days?

3 A radioactive source is measured to have an activity of 600 kBq. In 24 hours it has an activity of 75 kBq. What is the half-life of this source?

4 Describe how you would measure the half-life of a radioactive source in a laboratory.

In your description state:

(i) the apparatus you would use
(ii) the measurements you would take
(iii) how you would use the measurements to calculate the half-life.

5 State two safety precautions in handling radioactive sources.

6 State two ways in which you could reduce the equivalent dose of radiation reaching you.

7 State two places where you would see the sign for radioactive hazards.

15 Nuclear reactors

At the end of this chapter you should be able to:

1 State the advantages and disadvantages of using nuclear power for the generation of electricity.

2 Describe in simple terms the process of fission.

3 Explain in simple terms a chain reaction.

4 Describe the principles of the operation of a nuclear reactor in terms of fuel rods, moderator, control rods, coolant and containment vessel.

5 Describe the problems associated with the disposal and storage of nuclear waste.

Production of electricity

Electricity is the main source of energy in all countries, both for domestic and business use. While coal, oil, solar, gas, hydro and other sources of energy can be used for this purpose, they all must be able to generate electricity to be of use to society. The main reasons for this is that electrical energy can be changed easily into other forms of energy, particularly heat and movement, and electricity can be sent over large distances by a grid system.

However, the electricity must be produced at a power station, by one of a variety of methods. Renewable sources such as water, for hydro power, solar, wind and biomass such as wood will be able to generate electrical power. The amounts generated will be small in relation to overall demand but they are useful as additional supplies at peak periods. They have the advantage that these sources can be used almost forever without any problems.

For large-scale generation of power, power stations can use a variety of sources. These will all have to create heat. In the case of coal, oil and gas they are burnt to create the heat to turn water into steam. Nuclear power operates in a different way to produce the heat energy.

Advantages of nuclear energy
- No gases are produced which cause problems with the environment.
- While uranium is a limited resource there are sufficient quantities in the world to ensure large amounts of fuel.

- Oil may last for about a further 50 years and gas about 150, with coal possibly available for 300 years, but these fuels will become expensive to obtain. By comparison, a small amount of uranium, that is, a few kilograms, can supply several power stations.
- In the UK about 50% of energy is created from nuclear sources; in France it is about 70% and in the USA about 20%. These figures are likely to increase.

Figure 15.1 ▼
Radioactive waste

Disadvantages of nuclear energy

- The largest problem is the disposal of nuclear waste. Every power station will produce radioactive waste from the fuel rods. This waste has to be stored safely for hundreds of years. The amount of such waste is increasing (figure 15.1).

- There is always a danger of accidents. The official statistics show that nuclear energy is the safest form of energy production, but accidents can happen. The worst accident was at Chernobyl in Russia in 1986 (figure 15.2), when the workers ran the plant with the safety measures disconnected. The radiation from a damaged reactor was spread across a wide area of Europe by the wind and rain. Thirty-one people died from radiation-induced illnesses and 135,000 people living near the reactor had to leave the area. Overall the accident released 1.85×10^{18} Bq; 134 people showed signs of acute radiation sickness immediately after the accident.

Figure 15.2 ▶
Chernobyl after the accident in 1986

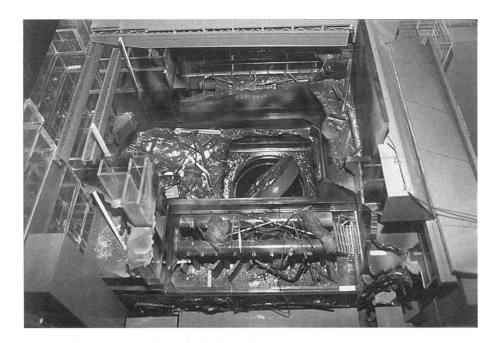

Fission

Nuclear power stations use uranium-235 as a fuel. To produce the energy required, the atoms of uranium are bombarded by neutrons. This makes the uranium unstable and it splits into two smaller pieces, as shown in figure 15.3. This process is called **fission**. The process releases energy and further neutrons.

Figure 15.3 ▶
Fission

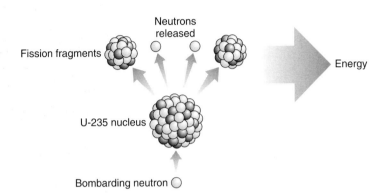

These neutrons can go on to hit further uranium nuclei, which split and again release energy and more neutrons. This process can continue and is called a **chain reaction** (figure 15.4). Such a reaction will take place in an atomic bomb and the amount of material needed to produce this is called the critical mass. In a nuclear plant the reaction must be controlled so that the amount of power produced is sufficient for our needs.

Figure 15.4 ▶
Chain reaction

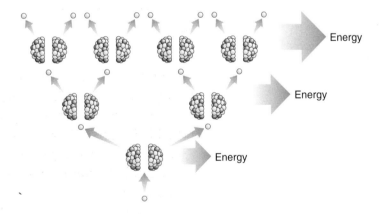

A nuclear power station

A nuclear reactor is illustrated figure 15.5. The key elements are as follows.

The fuel rods

These contain natural uranium which is enriched so that fission can occur. The uranium is found as uranium oxide and after being purified it is called 'yellowcake'. The amount of uranium in a fuel rod is well below the critical mass so that an explosion cannot naturally occur. The fuel rods will have to be replaced every few years. This is done by machine.

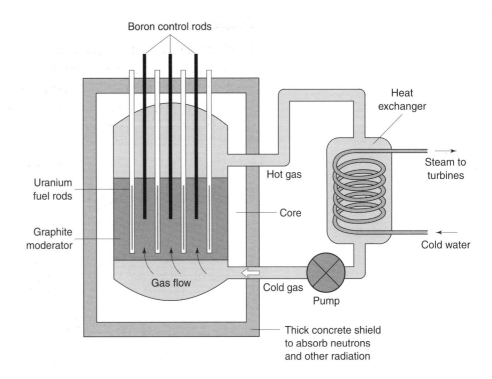

Figure 15.5 ▶
A nuclear reactor

Boron control rods

Heat exchanger

Steam to turbines

Hot gas

Uranium fuel rods

Core

Graphite moderator

Cold water

Gas flow

Cold gas

Pump

Thick concrete shield to absorb neutrons and other radiation

Graphite moderator

When the neutrons are emitted after fission they are moving very fast. They will not be able to be 'captured' by other nuclei so fission will not occur. However, if they are slowed down then there is a greater chance that fission will occur. This is done using a moderator. This is generally graphite. When the neutrons collide with the graphite atoms they slow down and can cause fission to take place in other uranium nuclei.

Control rods

The amount of electrical power required will vary, with peak demands during the day, whereas during the night the demand is lower. Rods of boron will absorb additional neutrons and control the number available for fission. They can be raised and lowered into the reactor as necessary. They also provide a safety feature since in the event of an accident all the control rods would be lowered automatically to absorb the neutrons.

Coolant

The heat produced during this process must be removed from the reactor. This is normally done in UK reactors using carbon dioxide gas. This circulates around the core of the reactor and is continually heated and then passes the heat to water by a heat exchanger. The water is not radioactive and can be reused. The water is then changed to steam, which is used to turn a turbine, which acts on the generators to produce electricity. This last stage is similar in any kind of power station after the heat is produced.

Containment vessel

The key parts of a nuclear reactor, which form the core, are contained in a containment vessel. This is designed so that no radiation can escape. It is generally several metres thick and has a concrete top. It is safe to walk on the top of the reactor but access is almost impossible unless it has to be repaired. The vessel can withstand a small conventional explosion.

Disposal of nuclear waste

Nuclear waste is split into different categories.

High-level waste is mainly spent nuclear fuel. After several years of use the fuel rods are taken out and sent to a reprocessing plant. These factories extract the useful parts of the fuel, which can be used to make new fuel rods. Unfortunately the material that remains is very highly radioactive. The materials are stored initially in a large tank of water. This absorbs the heat and also some of the radiation. After about a year the materials can be handled but are still very radioactive. The materials have very long half-lives and have to be stored in a suitable environment which is safe to human beings for a long time. Initially they were stored in large water-cooled tanks but these have further problems with leaks and cooling. The latest suggestion is to change the materials into a form of glass and then store it in boreholes deep in the ground. This has further problems since no one knows if the glass-like materials could crack, or if the radiation could leak into rocks and possibly into water supplies. This is still not considered the full solution and the waste is at present stored on site at the reprocessing plants. Part of the problem is the long time for the materials to be safe since scientists cannot guarantee storage of anything over a long period (figure 15.6).

Other suggestions have been to put the waste in a spacecraft and send it towards the sun – but what if the spacecraft explodes on launch?

Low-level waste is the remainder of nuclear wastes and those generated by hospitals etc. 'Low-level' does not mean it is not dangerous. It means that the radioactivity is less concentrated but the waste still has to be stored. It used to be dumped at sea but this is now banned since humans cannot continually pollute other parts of the environment. In a typical country, since nuclear reactors began in 1957 the nuclear waste amounts to about 9000 tons but the total amount of poisonous waste is about 50 million tons.

In summary, the storage of the waste centres on two issues:

- The disposal method; for example, the waste is made into a form of glass or stored in large tanks.
- The site for the storage.

Figure 15.6 ▼
Nuclear low-level waste storage facility

There is a tendency to think that nuclear waste is not our problem but someone else's problem. Since we all use electricity, some of which is nuclear generated, it is all our problem.

Physics facts and key equations for nuclear reactors

- There are advantages and disadvantages of using nuclear power for the generation of electricity.

- Fission occurs when certain nuclei are bombarded by neturons and become unstable. The nuclei break into smaller pieces and release energy.

- A chain reaction occurs when neutrons released by fission go on to strike more nuclei and cause more nuclei to split.

- A nuclear reactor has fuel rods, a moderator, control rods, coolant and a containment vessel.

- The disposal and storage of nuclear waste are difficult due to the half-life of some sources and the long-term storage that is required.

End-of-Chapter Questions

1 State two advantages and two disadvantages in using nuclear power to generate electricity.

2 Fission occurs when a neutron hits certain nuclei. Describe what happens next.

3 Explain why energy is released in a fission reaction by including in your answer an explanation of a chain reaction.

4 A nuclear reactor has several parts. Explain what is the function of each part listed below;

   fuel rods   moderator   control rod   coolant

5 Several sites have been proposed which can be used to store nuclear waste. State two reasons why the disposal of nuclear waste can cause problems.

Exam Questions

1 A radioactive source emits radiation with a half-life of one year.

(a) Explain what is meant by half-life.

(b) The source has an activity of 1 MBq. What will be its activity in 6 years?

(c) When radiation is used for treating humans, the effect of the radiation will depend on what three factors?

(d) Radiation of quality factor 20 has an energy of 0.35 J. It is absorbed by a person of mass 70 kg.

  (i) Calculate the absorbed dose

 (ii) The dose equivalent.

2 Nuclear reactors are the main method by which electricity is produced in the UK.

(a) Describe two advantages and two disadvantages of using nuclear energy.

(b) The production of this energy depends on **nuclear fission** and a **controlled chain reaction**. Explain what is meant by these terms.

(c) Radioactive waste is a problem with reactors. Describe some of the key issues in the disposal of this waste.

1 The graph below represents the motion of a cyclist travelling between two sets of traffic lights.

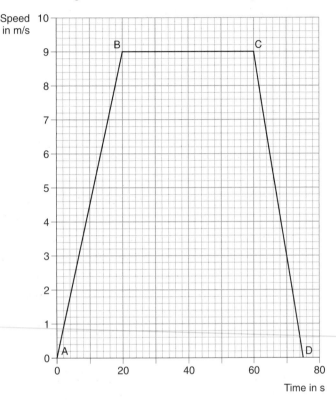

(a) Describe the motion of the cyclist
  (i) between B and C
  (ii) between C and D. **2**
(b) Calculate the acceleration between A and B. **2**
(c) Calculate the distance between the two sets of traffic lights. **3**
(d) Later in the journey the cyclist free-wheels down a hill at constant speed.
  Explain this motion in terms of the forces acting on the cyclist. **2**
  (9)

2 An aircraft magazine gives the following information about Concorde.

Mass	185 000 kg
Maximum speed	605 m/s
Take-off speed	112 m/s
Landing speed	83 m/s
Number of engines	4
Force from each engine	170 kN

(a) (i) Calculate the total force exerted by the engines.
  (ii) With all of its engines on, at one point on the runway Concorde has an acceleration of 3.20 m/s^2.
    Calculate the frictional force acting on Concorde at this point. **4**
(b) Some modern aircraft are controlled by on-board computer systems linked by optical fibres.
  The diagram below shows a ray of light inside an optical fibre.

  (i) (A) Copy and complete the diagram to show the path of the ray of light in the fibre.
    (B) Name the effect that occurs when the ray hits the inside surface of the fibre.
  (ii) In an optical fibre link in an aircraft, light travels a distance of 62 m.
    Calculate the time taken for a light signal to travel along this fibre. (speed of light in fibre = 2.0×10^8 m/s) **4**
    (8)

3 A chairlift at a ski resort carries skiers through a vertical distance of 400 m.
(a) One of the skiers has a mass of 90.0 kg. What is the weight of this skier? **2**
(b) (i) The chairlift carries 3000 skiers of average mass 90.0 kg in one hour. What is the total gravitational potential energy gained by the skiers?
  (ii) The chairlift is powered by an electric motor which is 67.5% efficient.
    Calculate the input power to the motor. **5**
    (7)

4 Some smoke alarms use a radioactive source which emits α-particles.
The detector operates because of ionisation caused by the α-particles in the space between the α source and the sensor.
If smoke or dust enters the space, the alarm operates.

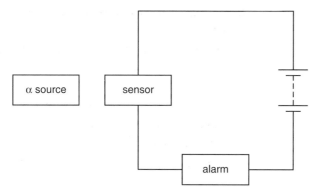

(a) (i) What is an α-particle?

   (ii) What is mean by "ionisation"?

   (iii) Why is an α source used instead of a source which emits β-particles or γ-rays?

The smoke alarm manufacturer has to choose from three α emitting sources. The half-life of each source is shown in the table.

Source	Half-life
A	4 hours
B	4 weeks
C	400 years

Which source should the manufacturer choose?   **5**
Explain your answer.

(b) One design of smoke detector has an LED which lights to show that the battery is in good condition.

A 9.0 V battery is used in the LED circuit shown below.

One component is missing, between A and B.

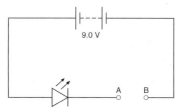

In normal operation, the LED carries a current of 20 mA and the voltage across it is 1.9 V.

(i) What is the name of the component that should be connected between A and B?

(ii) Calculate the value which this component should have so that the LED operates normally.   **4**
   (9)

5 Typical wavelengths in air of light of different colours are given in the table below.

Colour	Wavelength in air in m
red	6.5×10^{-7}
green	5.2×10^{-7}
blue	4.0×10^{-7}

(a) What is the speed of light in air?   **1**

(b) The frequency of a certain colour of light is 4.6×10^{14} Hz.
What colour is this light?
You must justify your answer by calculation.   **3**
   (4)

6 A converging lens has a focal length of 30 mm.

(a) Calculate the power of this lens.   **2**

(b) (i) In the diagram below, which is drawn to scale, an object is shown at a distance of 80 mm from a lens.
The points marked F are one focal length from the lens.

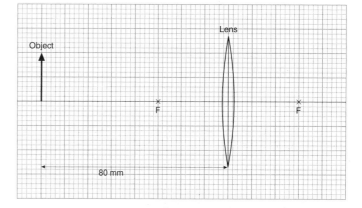

Copy the diagram onto squared paper and draw in two rays to show the formation of the image of this object.

(ii) State **two** ways in which this image is different from the object.   **5**

(c) This lens is now fitted to a camera. The next diagram next shows rays of light coming to the lens from a distant object.

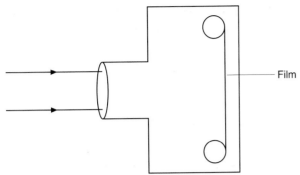

How far should the film be from the lens to give a sharp image?
Explain your answer. **2**

(9)

7 A drawing of the core of a nuclear reactor is shown below.

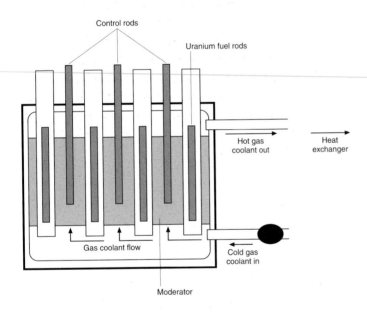

The fuel rods contain uranium-235.
(a) Describe what happens when a slow (thermal) neutron is absorbed by the nucleus of an atom of uranium-235. **2**
(b) The control rods are raised out of the core slightly.
Explain the effect of this action on the temperature of the coolant gas leaving the core of the reactor. **3**
(c) A research scientist at a nuclear reactor has a mass of 70.0 kg.
The scientist receives a dose equivalent of 336 μSv due to slow neutrons.
The energy absorbed by the scientist from the

neutrons is 8.40×10^{-3} J. Calculate the quality factor for slow neutrons. **3**

(8)

8 A lightning conductor is fitted to a tall building.

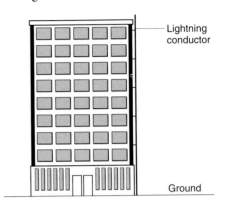

The specification for the lightning conductor is:
length   50.0 m
resistance per metre   0.080 Ω
mass   100 kg
specific heat capacity   385 J/kg°C.

During a thunderstorm, a total charge of 300 C flows through the lightning conductor to the ground in 0.120 s.
(a) Calculate the current in the lightning conductor during this time. **2**
(b) Show that the power in the lightning conductor is 25 MW. **3**
(c) (i) Calculate the maximum temperature rise that could be produced in the lightning conductor by this flow of charge.
(ii) What assumption have you made in your calculation (c)(i)? **4**

(9)

9 A student is given a piece of resistance wire 200 mm long and is asked to find its resistance. Part of the circuit the student builds is shown below.

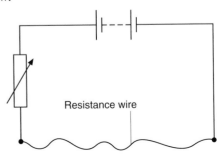

The student is also provided with a voltmeter and an ammeter.

(a) Redraw the diagram to show how the student should connect the meters to measure the resistance of the wire. **2**

(b) The student now uses measurements from the experiment to draw the following graph.

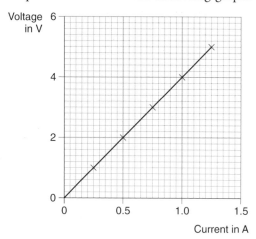

(i) Describe how the student uses the circuit to obtain the measurements for the graph.

(ii) Calculate the resistance of **one metre** of the wire. **5**

(c) Two pieces of wire have resistances of 2.0 Ω and 6.0 Ω.

The wires are connected together as shown below.

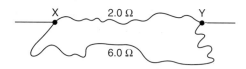

Calculate the resistance between points X and Y. **2**

(9)

10 The circuit shown below is used to investigate the behaviour of component X.

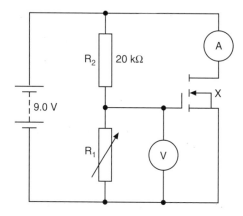

(a) (i) What is the name of component X?

(ii) What name is given to the arrangement of the two series resistors R_1 and R_2 connected across the 9.0 V supply? **2**

(b) By using different values of resistor R_1, different voltages are applied to the input of X. The graph below shows how the current through X changes as the voltage applied to the input is altered.

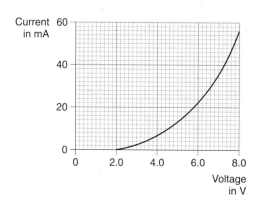

(i) What is the reading on the voltmeter when device X starts to conduct?

(ii) Calculate the value of resistor R_1 which is required to obtain this voltage. **3**

(c) Component X is now connected into the circuit below to switch on an emergency lamp when it becomes dark.

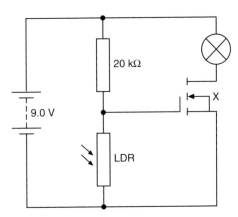

The table below shows the resistance of the LDR in light and in dark.

Lighting condition	Resistance of LDR in kΩ
light	0.1
dark	10.0

Explain how this circuit operates to switch on the emergency lamp.

3

(8)

Data Sheet

Speed of light in materials

Material	Speed in m/s
Air	3.0×10^8
Carbon dioxide	3.0×10^8
Diamond	1.2×10^8
Glass	3.0×10^8
Glycerol	2.1×10^8
Water	2.3×10^8

Gravitational field strengths

	Gravitational field strength on the surface (N/kg)
Earth	10
Jupiter	26
Mars	4
Mercury	4
Moon	1.6
Neptune	12
Saturn	11
Sun	270
Venus	9

Specific latent heat of fusion of materials

Material	Specific latent heat of fusion (J/kg)
Alcohol	0.99×10^5
Aluminium	3.95×10^5
Carbon dioxide	1.80×10^5
Copper	2.05×10^5
Glycerol	1.81×10^5
Lead	0.25×10^5
Water	3.34×10^5

Specific latent heat of vaporisation of materials

Material	Specific latent heat of vaporisation (J/kg)
Alcohol	11.2×10^5
Carbon dioxide	3.77×10^5
Glycerol	8.30×10^5
Turpentine	2.90×10^5
Water	22.6×10^5

Speed of sound in materials

Material	Speed (m/s)
Aluminium	5200
Air	340
Bone	3000
Carbon dioxide	270
Glycerol	1900
Muscle	1600
Steel	5200
Tissue	1500
Water	1500

Specific heat capacity of materials

Material	Specific heat capacity (J/kg°C)
Alcohol	2350
Aluminium	902
Copper	386
Diamond	530
Glass	500
Glycerol	2400
Ice	2100
Lead	128
Water	4180

Melting and boiling points of materials

Material	Melting point (°C)	Boiling point (°C)
Alcohol	−98	65
Aluminium	660	2470
Copper	1077	2567
Glycerol	18	290
Lead	328	1737
Turpentine	−10	156

SI prefixes and multiplication factors

Prefix	Symbol	Factor	
mega	M	1 000 000	$= 10^6$
kilo	k	1000	$= 10^3$
milli	m	0.001	$= 10^{-3}$
micro	μ	0.000 001	$= 10^{-6}$

Index